T0065225

THE FATE OF
A.I. SOCIETY

THE FATE OF
AI SOCIETY

Civilizing Superhuman Cyberspace

Kenneth James Hamer-Hodges

Archway Publishing books may be ordered through booksellers or by contacting:

Archway Publishing
1663 Liberty Drive
Bloomington, IN 47403
www.archwaypublishing.com
844-669-3957

ISBN: 978-1-6657-4972-5 (sc)
ISBN: 978-1-6657-4971-8 (e)

Library of Congress Control Number: 2023917133

Print information available on the last page.

Archway Publishing rev. date: 09/25/2023

DEDICATION

To Christine,

You are the rock of our family and the source of my inspiration. Your steadfast support and love are the foundation to everything we achieve together. Your strength, skill, and resilience are my guiding lights. Importantly you were the ultimate editor of this book. It is your unwavering faith that gives me courage to pursue this dream.

Thank you for your love, your patience, and your resolute support as the heart and soul of our family and our friendships. I am blessed to have you by my side as we journey through the unknown.

With all my love and gratitude for guiding my happy and rewarding life...

The world today is made; it is powered by science, and for any man to abdicate an interest in science is to walk with open eyes towards slavery.

—J. Bronowski, from *"Science and Human Values,"* when at the Salk Institute for Biological Studies, San Diego, California, February 1964

CONTENTS

LIST OF FIGURES

INTRODUCTION

The phrase *"when good men do nothing"* is attributed to the Irish statesman and philosopher Edmund Burke. The full quote from *"Thoughts on the Cause of the Present Discontents"* (1770) as the American Revolution was fermenting goes: *"When bad men combine, the good must associate; else they will fall, one by one, an unpitied sacrifice in a contemptible struggle."*

This quote highlights that morally upright individuals must take action in the face of injustice, oppression, or wrongdoing. Remaining passive or indifferent in such situations allows harmful actions to persist and worsen. The sentiment behind this phrase encourages individuals to stand up for what is right, even if it's uncomfortable or challenging. It emphasizes the individual responsibility of citizens to contribute positively to their society and not let immoral or harmful actions go unchecked.

The combination of Artificial Intelligence (AI) with first-generation binary computers and a frozen hardware industry presents such a moment. Binary computers, selfishly frozen in the past by the computer industry, strand humanity with unanswered software threats. Over the past five decades, code hacks that began as gimmicks became malware for criminals, spies, and weapons in international cyber wars. It will crush individuals, businesses, and the nation. Worse still, the present defense only increases the constraints on individuals, democracy, and freedom. The incremental growth of digital dictators is Orwellian but as crazy as Lewis Carol's dreams invented for Alice.

Not for the first time in the history of computer science, discarding decades of hard work is disappointing, but once again, it is the right choice. When Henry Briggs explained to John Napier that he should pitch twenty years of hard work on logarithms to using the base number ten, it was not easy, but he agreed. Why? Because everyone gained by democratizing logarithms for the public at large.

Orwellian AI-powered dictatorship is the threat. Discarding half a century of work on binary computers will be more challenging but essential. Everyone will gain if an AI-enabled society is democratic, safe, and equally so for all.

BIG BROTHER

When Good Men Do Nothing

The novel *1984* by George Orwell depicts a dystopian society in which the ruling party led by Big Brother has complete control over every aspect of citizens' lives. They must show unwavering loyalty and devotion to him. Big Brother is a metaphor for the state's power and the dangers of totalitarianism. The expression *"Big Brother is watching you"* has become a cultural reference to oppressive governments and surveillance states, even democratic ones. The rise of dangerous digital technology, street cameras, the dark side of the internet, deep fakes, facial recognition, and AI makes deploying Big Brother easy. Governments, helped by greedy industries selling their latest inventions, are happy to do so.

This tragedy is well-advanced in China, Russia, Iran, and North Korea. Furthermore, international criminal gangs in Eastern Europe and the Far East, industrial dictators like Google, Microsoft, Intel, and others, and government inaction erode freedom, equality, and justice in cyberspace. They break the critical cornerstones of trust in democracy. The dark forces of malware act as Alice's Red Queen, from Lewis Carol's Wonderland, at work, manipulating information to influence individuals, even tinkering with the large language models to redirect AI malevolently.

The core problem is the centralization of power in computer science, and individuals must resist it. Citizens need rights to protect themselves and democracy from dictatorship, but no such individual powers exist in cyberspace because digital privacy, the first essential, is missing. Democracy depends on the votes of free, equal, private, and independent citizens. But as a dictatorship, binary computers suppress human rights, as enforced by the superuser architecture of the centralized operating system in every binary computer. As cyberspace strengthens its grip on life, it blocks the freedom and individuality needed to sustain democracy. Consider the claim of election fraud and corruption surrounding the US presidential elections.

Even though the Cybersecurity and Infrastructure Security Agency, responsible for election security, called the 2020 election *"the most secure in American history,"*[1] not everyone agreed. Lawsuits alleged voter fraud and other irregularities. Judged to lack evidence or standing, they went nowhere, leaving unhappy citizens without options. Public discontent expressed on social media is the problem of truth and lies. It is still an open problem, and cyber society must deal with the insidious impact amplified by outsiders and foreigners wishing to stir up internal troubles. The opaque nature of binary computers and the lack of transparency with the increasing use of superhuman AI amplifies the problems. Traceability is an effective deterrent that, like security, must work from the edge of computer science to catch every change. In every case, the science of nature addresses these problems. Life needs no central power; it is individuality that sustains society.

Urgently, the next generation of computer science must address these issues. The Church-Turing thesis[2], the cornerstone of computer science, explains the answer. Computer science, like mathematics, extends the power of individuals using private threads. Privacy replaces centralized dictatorship by following a functional implementation of the λ-calculus.[3] Software modularity is protected and guaranteed

---

[1] Cybersecurity experts say public-private partnership is the key; https://valapp. dynu.net/2021/11/16/cybersecurity-future-government-corporation-partnership-data-breach/.

[2] Church-Turing thesis, Wikipedia states that a function on natural numbers if it is computable by humans, a Turing machine, and the λ-calculus, named after American and British mathematicians Alonzo Church and Alan Turing.

[3] Lambda calculas, Wikipedia. Lambda calculus (also called λ-calculus) is a formal system in mathematical logic for expressing computation based on function abstraction and application using variable binding and substitution. It is a universal model of computation and can simulate any Turing machine.

by the digital mechanics of Capability-Based addressing[4] framing object-oriented machine code. If not, things will only get worse. Crime and polarization will lead to dictatorship and Big Brother, all enhanced by superhuman AI.

Consider that in October 2020, weeks before the presidential election, Twitter blocked news about the son of a candidate. Twitter executives suppressed articles, citing a company policy against *"hacked materials,"* only to admit months later this was wrong. Decisions to block information amount to censorship and bias. A story buried by any dictator in cyberspace conflicts with the free speech amendment to the constitution. However, the systemic centralized powers in binary computers prevent citizens from responding to the misuse of centralized authority. Centralization has limited an individual's ability to protest. Industrial dictators already overrule equality and civil justice in cyberspace.

For democracy to function, power must emanate from individuals, from the hands of *"We the People."* Power cannot pass through any dictatorial operating system. Civilizing cyberspace demands the scientific replacement of centralized software systems using the mathematics of Alonzo Church's λ-calculus. Decomposition into λ-calculus function abstractions is the scientific solution for individuals in cyberspace. Capability-based addressing protects the named structure using symbolic tokens defined by a λ-calculus namespace as individual, independent functional applications. The immutable digital tokens enable secrecy, privacy, and individuality across cyberspace as golden digital assets.

Placing cyber power with the people is achieved through the immutable tokens by which the λ-calculus frames all computations. Tokens channel AI and tame malware while, at the same time, forcefully expressing the foundations of individuality and democracy. It begins with the equality of science as a flawless digital implementation. The

---

[4] Capability-based addressing limits access to memory as segments with type limited and range limited boundaries. Immutable pointers replace physical addressing called capabilities created through secretly hidden function abstractions. Thus, the system limits access to the minimum necessary portions of memory (and disable write access where appropriate), without needing separate address spaces and switching context. The PP250 used capability pointers to address local and network functions objects. https://en.wikipedia.org/wiki/Capability-based_addressing.

tokenized cyberspace defined by the Church-Turing thesis corrects every unsolved binary problem. With this critically important change, centralization disappears as citizens hold power through immutable namespace tokens representing the incremental functionality of the λ-calculus. It reimplements the Church-Turing thesis as a dream of flawless cyberspace. Now, power is distributed incrementally through the hands of *"We the People,"* privacy is automatic, security is individual, and unique computations belong to individual citizens. Privacy allows individuals to struggle against digital oppression and fight to redefine democracy using AI-enabled cyberspace.

It is still vital for social media companies to set company policies about sharing on their platforms as private companies. At the same time, the First Amendment of the US Constitution that protects freedom of speech must apply to cyberspace. However, cyberspace belongs to us all. It is the abstraction of science, knowledge that belongs to everyone. Private companies can set rules and policies; users should abide by them if they use the service.

Nevertheless, there are legal and ethical considerations around the role of computers and social media in shaping public discourse and the free flow of information. Indeed, as a public service and a fact of nature, computer science extends individuals, not the centralized power of binary dictators or AI-enabled malware. It makes freedom in cyberspace a human rights concern.

While private companies have the right to set internal policies, they must still act in the public interest. Unfair actions that change the public debate or the democratic process are undemocratic. As a global public service, cyberspace must be neutral to all, but it is not so today. These conflicts demand extensive debate because digital dictators play significant and expanding roles in shaping national conversation and must account for their actions. For a constitutional republic, cyberspace as an extension of *"We the People"* is the extension of democratic government. The Constitution must define this new relationship. Clarification means a constitutional debate, resulting in any amendment made *"by the people for the people"* on citizens' rights, digital security, and computational privacy in cyberspace.

The outdated, centralized, overstretched, and overshared binary computer threatens a happy, democratic life in a civilized cyber society. At the same time, science is the neutral party that will save democracy by handing power to the people. The uniformity and equality of mathematics in the Church-Turing thesis is the scientific flawless, fail-safe platform for cyberspace. The lesser binary alternative used today is forever tainted and unacceptable.

The nature of bias in corrupt dictatorial computers is inscrutable, and global cyberspace makes purging Big Brother and corruption from the network impossible. AI malware will exist forever, but concerns over digital dictators and undetected corruption will disappear by replacing binary computers with Church-Turing machines. And therefore, most importantly, it is how to empower the people by preventing malware, identifying crime, including deep fakes, finding digital criminals, taming malware, and targeting AI. Using science, engineers can replace dictators with cyber democracy. The golden mechanics of tokenized individualism will prove the lifeblood of forceful, sustainable digital democracy.

In 1936, at Princeton University, computer science was researched and defined as the Church-Turing thesis, which recognized networked function abstractions, limitlessly bound to the science of mathematics through the λ-calculus. Software is ideal when defined and organized by the λ-calculus as individual, independent functions in namespace applications. The namespace and the function abstractions are separate objects, atomically programmed as logical functions in expressions.

Neither a programmed abstraction nor a namespace needs a centralized operating system. Instead, power is incremental, distributed secretly as golden Capability tokens, function by function following the need-to-know rule of top security systems. Individual programmers allocate the tokens on a need-to-know basis to the modular abstractions of the namespace. Individual human users, as citizens, are granted access tokens as private, immutable digital keys to unlock access to digital objects. Directional, hierarchal, programmatically approved authorities, each one, when necessary, is democratically authorized as a *"chain of office."* Immutable tokens are the golden fabric of cyber democracy, backed by access rights to the roles of software as a nation of laws.

Only pure computer science can protect the future of individuals and society. The pure science of the λ-calculus changes the nature of cyberspace to the fail-safe, flawless, level playing field of

Church-Turing machines. The binary computer is just a Mad Hatter, a weird human concoction constantly changing the rules as guests try to work. Democratizing cyberspace requires centralization to disappear.

But the design of a branded binary computer tilts everything towards undetected corruption, centralized dictators, skilled hackers, remote attackers, international criminal gangs, and enemies of the state. Alternatively, the laws of mathematical science level the playing field of cyberspace into a flawless, neutral, egalitarian, efficient, and secure approach to functional, digital computers. Private function abstractions, accessible only by unique, private tokens owned by individuals, hide, limit, and secure access to even the most sensitive atomic data. These immutable digital tokens are the gold of digital security. The secret belongs to the owner, preventing access to all others, including malware, leveling the field of play as defined by the λ-calculus.

Unquestioningly, trusting programmers to follow best practices and freely access linearly shared memory will remain a catastrophic digital mistake. Digital corruption caused by accident from a program bug or intentionally by a criminal attack remains unnoticed for weeks and months, allowing corruption to fester and grow. The dark motives of humanity win every time. Instead, the science of the λ-calculus constrains and targets AI society despite any dark forces using private, secret tokens of immutable digital gold.

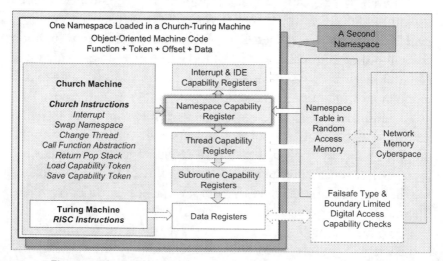

Figure 1. The encapsulated machine architecture for the λ-calculus

The only way to achieve democracy in cyberspace is to privately grant the functional power of λ-calculus tokens to the people using the digital keys that unlock the embedded access rights of Capability-based addressing.[5] Otherwise, undetected digital corruption, unpunished crime, and international conflict will continue to destroy democracy through digital dictatorship, propaganda, surveillance, and disinformation. Tragically, without question, society accepts centralized computer power run by industrial dictators plagued by criminals, spies, and enemies. These threats increase as children now grow to take this dangerous new world for granted, as AI simplifies the darkest sides of criminal cyberspace.

The entire world is digitally dependent and struggling to survive. Recharging smartphones is a top priority even for displaced immigrants disembarking from rubber boats after crossing the Mediterranean Sea or the Rio Grande. But computers that power this Brave New World[6] are dystopian, the binary foundations are flawed, and powerful software using AI is further misguided. Digital civilization cannot survive on this corrupt foundation; the independence needed for a functioning digital democracy is already dead in the centralized binary computer.

Individual freedom and equality cannot exist if industrial dictators, rampant crimes, and backward compatibility continue to stifle progress. Civilized nations waste increasing effort fixing flaws that engineers could prevent by default as respectable professionals. As the cracks in computerized society develop, law and order crumble, replaced by industrial dictators, criminal gangs, and enemies who wish to rule cyberspace. Like Big Brother, industrialists and criminals use fear to scare individuals into mistakes, lock home clients, limit progress, maximize revenue, and dominate their markets. These industrial dictators are replacing democracy. Their Orwellian vision of life under Big Brother is a corporate strategy. It must stop. Breaking up these industrial dictators will not fix these problems. They will reappear. Instead, backward compatibility that sustains centralization must end to allow free competition and progress,

---

[5] https://en.wikipedia.org/wiki/Capability-based_addressing.
[6] Brave New World, Wikipedia.

leading to open computation based on the equality of the λ-calculus. It is the only way democratic civilization will survive.

As cyberspace evolves, digital software takes over ever more control. Traditional life overrun by this synthetic alternative is crude. It must be scientific and thought out. The digital universe of confused, discontinuous, and conflicting machines ruled by almighty software, supervised by operating system monopolies, is constantly changing when attacked by criminals and enemies worldwide. Digital information, brought to life as virtual reality, reflects computer software, disgorged from an information matrix processed by ever more powerful semiconductors, controlled by AI and a few industrial potentates. The reconstruction of society drawn from foreign edges of the internet is a digital reflection paramount to the quality of national life as a virtualized democracy.

The integrity of software as virtual reality, or the metaverse, is fundamental to the survival of AI society. But digital corruption is dangerous, unnatural, and disastrously undemocratic. Software automation will subsume everything essential to the pursuit of happiness, from the digital economy to social media and from law and order to the progress of democracy. Flawless computer science is vital because AI society is contingent upon software integrity. To remove unnatural bias, achieve scientific excellence, resolve cybercrime, prevent cyberwar, and empower individuals and digital democracy, perfect computer science is critical.

For the future to work for all, as a progressing civilization, computers must place all software power equally, evenly, and democratically into the hands of the people, not the dictators. Only then is humanity guarded against foolish mistakes. Defanging dictators requires the golden tokens to decompose and decentralize, avoiding monolithic centralization. Abstracting the traditional locks and keys of the physical world using immutable gold as digital tokens with Capability Addressing is the way to software modularity. Functional modularity defines the laws of λ-calculus. Golden tokens work as private keys, which is how the λ-calculus, Capability-based addressing, and object-oriented software all work together. Distributing power as golden tokens into the hands of individuals, citizens, and roles governed by elections in a democratic society is vital. It must be so before it is too late.

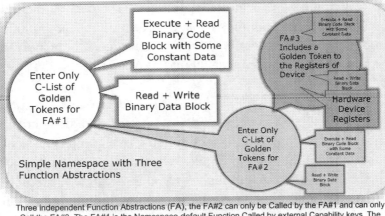

Three independent Function Abstractions (FA), the FA#2 can only be Called by the FA#1 and can only Call the FA#3. The FA#1 is the Namespace default Function Called by external Capability keys. The FA#3 can also access some hardware registers of an attached device, that could include the computer

Figure 2. The fail-safe Objects in a Church-Turing Namespace

The digital miracle of computer science is not subject to traditional law and order. It is presently designed by and for industrial dictators. The binary computer locks customers into a dictator's grip. Fear of Alice's Red Queen[7] and Orwell's Big Brother will overwhelm life. Ever-present, systemic cybercrime enables the uncomfortable domination of industrial dictators. This painful reality is unscientific. Mathematics, as the foundation of computer science, is flawless. Computer science should be the same as it was for Charles Babbage. He showed Ada Lovelace his perfect machine for mathematics in 1834, at the height of the Industrial Revolution. She dubbed his computer the *thinking machine*.

However, unlike Charles Babbage's *thinking machine*, binary computers are plagued by undetected, silent, endless crime. This systemic problem exists because binary computers are shared, unscientific, unreliable, and centralized, designed by the ego of engineers following in von Neumann's[8] footsteps as privileged

---

[7] Through the Looking Glass. The Red Queen is a fictional character and the main antagonist in Lewis Carroll's fantasy 1871 novel Through the Looking-Glass. Not to be confused with the Queen of Hearts from the previous book Alice's Adventures in Wonderland (1865). They are two are very different characters.

[8] John von Neumann is known known for his work in the early development of computers: As director of the Electronic Computer Project at Princeton's Institute for Advanced Study (1945-1955), he developed MANIAC (mathematical analyzer, numerical integrator, and computer), which was at the time the fastest computer of

superusers running mercenary dictatorships. The central operating system and the system administrator enforce the suppliers' solution. Without the operating system, binary computers have only the limited, stand-alone program capability as Alan Turing's first paper tape machine.

Today, industrialists define the laws of cyberspace, using proxy software to bend the rules to branded wishes. As in a democracy, it must be the other way round; society must rule computer science, and the λ-calculus with Capability-based addressing and object-oriented programs allow this goal to succeed. Inevitably, corruption flourishes under dictators, and as law-and-order fails, only chaos and human suffering remain. This tragic endgame is not inevitable. Like mathematics, precise computer science is crime-free, guarding the citizen against mistakes, like Babbage's *thinking machine*, governed scientifically and democratically equal for all. When operated as an extension of free and independent individuals, democracy is improved by AI cyberspace.

Cyberwars rage to dominate cyberspace. Industrialists protect their outdated binary design, others build software empires, while hackers, gangs, and enemies are spoilers. They all feather their nests at the expense of citizens. Life under any dictator is wrong. Thus, binary computers as digital dictators are unacceptable because they undermine future life, the pursuit of happiness, democracy, and civilization. Unless individuals stand up and demand changes that prevent undetected digital corruption and decentralization of dictatorial operating systems in computer science, democracy will end catastrophically.

When computers began, experts expected the programs they wrote themselves to work. John von Neumann, the self-proclaimed architect of general-purpose computers, corrected his machine code in his head. He was impatient, witty, and incalculably brilliant. Industry experts quickly accepted his proposal, prematurely published under his name to deliberately sidestep patents and claim the invention in his name. The computer industry now enjoys an unfair advantage over society. It must end. Virtual memory and

---

its kind. He also made important contributions in the fields of mathematical logic, the foundations of quantum mechanics, economics, and game theory.

centralized operating systems must go. Instead, the level playing field of the Church-Turing thesis that avoids privileged hardware and superusers must dominate.

The cause of malware and cybercrime is that no one can trust globally connected binary software. Once locked in a room, guarded by the owners, disconnected from the world, blind trust worked for the cold war days of batch processing, but the fatal details passed onto the personal computer. Information technology departments, the feared IT mafia, grew to enforce the Cold War standards and prevent networked interference. In addition, they evaluated every line of code for months before approving software on the corporate mainframe. But, individual users of the PC remain ignored by industrial dictators.

Cybercrime exploded once hackers discovered the dangerous default authority gifted by the binary hardware to every program. Crooks quickly followed hackers to interfere with the stability of the software for financial gain. Without difficulty, they spy on and steal personal data from outdated binary computers. The digital void exists by default, launching the industry on a path of endless patched upgrades. It is a race they always lose, trying to guess and block the next zero-day attack[9].

Sadly, binary computers will always exhibit fatal flaws. The digital voids are virtual machines enslaved to an operating system. Systemic flaws like the confused deputy attack[10] always succeed, limited only by increasing complexity. Deep inside the NSA, the rule says the only safe computers are disconnected or, better yet, switched off. Practically speaking, neither alternative helps society. Computers offer tremendous value to everyone. When perfected, mathematically flawless computer science is vital to individuals living in the future.

By way of validation, consider that mathematics needs no centralized dictator, is universally equal to all, and is scientifically flawless. For example, perfect mechanical computers culminated with Babbage's *thinking machines*. Each computer, starting with the

---

[9] Zero-day attack, Wikipedia.

[10] The Confused Deputy Problem is a computer program tricked by a program with user privileges and less rights into misusing its authority. It is a specific type of privilege escalation and as the first example it is often cited as why capability-based security is important.

abacus and later following the slide rule, served the public interest, proven by the burst of civilized progress they achieved. They foster everyone equally, without interference. Unlike the flawed, opaque human concoction of outdated binary computers, pure computer science is mathematics and logic. The guardrails of science safeguard innocent, unskilled citizens in every industrial society by preventing mistakes and mechanizing reliable results controlled by any interested public.

Industry dictators and authoritarian governments use unfair computer privileges to dominate life and spy on citizens. They will argue that the present situation is good enough, but it is not so. When computers run the world, the overwhelming power of software will enslave society in binary dictatorships. The self-interests of industry, governments, and criminals use artificial intelligence to their ends while citizens and businesses suffer. Opaque powers are at war with one another to dominate cyberspace. These self-interested forces use the dictatorial nature of binary computers and crime in cyberspace to enforce their rules. They will build deep fakes[11] of famous individuals representing a new form of Big Brother, and innocent citizens will have no chance.

Governments have acted to improve computer security, protect citizens' data, and combat cyber threats. Examples include:

1. Legislation to ensure organizations protect sensitive data and take measures to prevent cyberattacks, like the GDPR in Europe and the California Consumer Privacy Act (CCPA) in the USA.
2. Cybersecurity strategies to identify and limit threats when sharing and collaborating between government agencies, private sector partners, and the public using education and awareness campaigns.
3. Cybersecurity centers are where expert resources can help individuals and victims of cyberattacks, serving as a point of contact, an information source, and coordinating any cybersecurity response.

---

[11] Deepfake, Wikipedia. Deepfakes are synthetic media in which a person in an existing image or video is replaced with someone else's likeness and says things out of context.

4. The United Nations Group of Governmental Experts on Information and Telecommunications exchange norms, principles, and rules of behavior in cyberspace offering international advice.
5. Law enforcement agencies that investigate and prosecute cybercriminals.

But no one considers the main danger of centralized dictatorship because the industrialists already have a stranglehold on computer science. While government actions to improve computer security and protect citizens from cyber threats help, they do not solve the problem of frozen hardware architecture from the Cold War. This outdated form of computer science lacks essential hardware updates, leaving software threats to grow in frequency and scope encouraged by AI-breakout, increasing the competence of undetected malware and global crime to superhuman complexity.

Critically, backward compatibility must end. Unfreezing hardware progress and ending backward compatibility is vital to keep pace with AI-empowered cybercriminals. A small step in the right direction has occurred at ARM[12] with the rediscovery of Capability-based addressing. But, until the dictatorial operating system and centralization disappear, all the existing threats to the future of nations and civilizations remain.

---

[12] Arm Morello Program https://www.arm.com/architecture/cpu/morello

DIGITAL CONVERGENCE

First-Generation Binary Computers

Digital convergence created the convenient but existentially dangerous global medium of binary cyberspace. Software in a binary computer lacks any private physical structure. Compiled as a monolithic binary image, the functional boundaries are blurred. This first-generation view of modularity lacks the critical boundary details to prevent interference. These virtual machines defined for batch processing decades ago uncomfortably share cyberspace. Consequently, cyberspace is unsafe, favoring dictators, enemies, and criminals using malware, spyware, ransomware, and AI to disrupt and dominate society. Citizens are overwhelmed by unnecessary complexity while suppliers deliberately sustain this outdated and opaque complexity to retain their captive markets.

Instead, the frozen hardware of the Cold War must catch up with AI-enabled software for democracy to survive. In AI cyberspace, computers must extend the power of individuals, not increase the strength of dictators who run binary computers as privileged superuser playgrounds. Cyberspace must power individuals to regulate democracy with equal intensity for all. As defined by science and nature, the laws of mathematics and logic are the cornerstones of civilization. It began with the abacus in Babylon and progressed to the infallible logic recognized in 1936 as the Church-Turing thesis.

But selfishly, suppliers froze the binary computer decades before the internet using the first-generation architecture of John von Neumann. This overstretched, over-shared binary computer needs a central operating system as a dictatorial superuser enslaving monolithic binary compilations, leaving users exposed, deprived, and dysfunctional. Hackers, spies, criminal gangs, industrial dictators, governments, and enemies capitalize on this unfair disparity to exploit the scientific flaws as digital gaps in the mechanics of cyberspace. Enslaved users are powerless and ignored; they cannot resist, so, in frustration, they dangerously expose the power of the superuser to crime and ransomware attacks.

Outdated by the passing decades, each brand of binary computer demands ever-increasing hours of exceptionally skilled human effort to fix malware and keep programs running. The deliberately opaque, digitally unsafe, and scientifically flawed binary computer is the enemy of progress and the stability of a sustainable AI-enabled society. Antiquated by time and besieged by pervasive, globally propelled malware, the backward-compatible binary computer is obsolete. Distracted by attacks, industrial dictators struggle to survive, distributing monthly upgrades too often ignored and subverted by hurried and unscientific decisions from a past that caught up. In global hands, the tidal wave of AI will overwhelm the world.

Governments, suppliers, spies, criminals, and enemies already have the upper hand on this tilted, binary playing field. The guardrails of physical, engineered designs built to pass the tests of time by detecting and constraining malware are missing. Centralized software in shared binary computers, frozen in the 1960s, lacks the structure defined for computer science by the λ-calculus expressed in the Church-Turing thesis when digital computers began. Binary computers cannot resist corruption because the networked science of λ-calculus is missing.

Alonzo Church[13] and his graduate student Alan Turing[14] defined two sides to the fair distribution of computational power. The

---

[13] Alonzo Church (June 14, 1903 – August 11, 1995) was an eminent US mathematician and logician with works of major importance in mathematical logic, recursion theory, and theoretical computer science. He is best known for the lambda calculus, Church-Turing thesis, proving the undecidability of the Entscheidungsproblem, Frege–Church ontology, and the Church–Rosser theorem.
[14] Alan Turing, a visionary mathematician, computer scientist, and logician, widely regarded as the father of modern computing and artificial intelligence. His groundbreaking work during World War II at Bletchley Park broke the German Enigma code at a pivotal moment in history, turning the tide of the war at sea and on land. At King's College, Cambridge, he studied mathematics,

combination works top to bottom by encapsulating Turing's binary computer as the lambda engine of Church's λ-calculus. In harness, they provide computational concurrency as secure, networked, corruption-free, private computations. Called the Church-Turing thesis, it is the cornerstone of computer science. Understanding this modular mathematical science, ignored in the heyday of isolated mainframes, is essential for computer science to unfreeze hardware progress, prevent malware, and constrain the power of AI as extensions of democracy. Atomic modularity is nature's form of universal law and order. Empowering equality for individuals is the law for human democracy, and this avoids dictatorships that crush humanity. Software modularity purges digital interference and enables individuality throughout cyberspace by removing the centralization of operating dictatorships.

The computational weakness of centralized, autocratic operating systems cannot cope with ever-more dangerous criminal attacks and the avalanche of superhuman threats created by AI, interconnecting criminals and enemies worldwide. AI-enabled software has powerful benefits, but power corrupts, and AI-enabled society demands cyberspace evolves to cope with this overwhelming, everlasting, existential condition.

People's power sustains democracy when they take to the streets. But no roads exist in binary cyberspace; power is hidden and opaquely dictatorial. These centralized dictatorships that crush democracy must disappear. The trade routes and roads connecting individuals as a digital civilization of equals must be transparent and malware-resistant, or George Orwell's threat of Big Brother will rule worldwide.

China, Russia, Iran, and North Korea lead the way. Democracy is easily lost and quickly subverted by the centralized dictatorships

---

and afterwards researched the Entscheidungsproblem with Alonzo Church at Princeton University in the USA. His proposal of the α-machine, now called the Turing machine was part of his research with Alonzo Church. After earning his Ph.D. Turing returned to England by 1938 to work on the war effort. Tragically, on June 7, 1954, Turing committed suicide at the age of forty-one, attributed to the drug 'treatment' he received to suppress homosexuality. His groundbreaking work received the recognition it deserved, when in 2009, the British government issued an official apology for the treatment of Turing while acknowledging his inspirational contributions to science.

of binary computers. The added power of AI to binary computers will extinguish democratic laws and order. AI will then destroy cultures and communities, just like the first Gothic cathedrals[15] that lacked structural engineering, killing or harming everyone in an innocent community who shared the place of worship. An AI breakout in binary cyberspace creates a Weapon of Mass destruction. This WMD could potentially end civilized society and industrial nations simultaneously and worldwide.

Hardware frozen in the past cannot serve society. Hardware must stay abreast of software to level the AI playing field and empower individuals instead of organizations by preventing dictatorships. Then, advanced malware is constrained by the laws of science, as expressed by the missing half of the Church-Turing thesis. Even AI malware that learns to lie cannot break the rules of logic and mathematics. Individuals and the laws of nature dictate the fate of AI society.

Advanced software is the Red Queen in Alice's mystical chess game, with powers to cheat, ignore, and usurp the rules of binary computers, unfairly using malware to redefine winners and losers throughout cyberspace. Computer hardware and software must be scientifically neutral, flawlessly obeying science. But in a centralized, dictatorial binary computer, AI becomes a WMD until scientifically constrained as an extension of incremental powers managed by independent golden tokens. The breakout of superhuman AI will enslave society as nations become digital dictatorships, ruling the world in ways beyond the interests of humanity.

Only dictators, criminals, and enemies confidently misuse the global reach of such dangerous binary powers. Improved attacks on individuals, businesses, and governments swiftly steal confidential data. They overwhelm opinions on social networks, forge news and identities, scramble data and results, overrule democratic law and order, demand money, and turn democracy into a digital dictatorship. Uncivilized terrorists, greedy industries, skilled hackers, and motivated hacktivists will break down society, making life unbearable, far worse than the Wild West, without any hope of individual salvation.

---

[15] Beauvais Cathedral https://en.wikipedia.org/wiki/Beauvais_Cathedral

Ransomware attacks highlight the failure of binary computers and centralized operating systems. As functional power splits between good and evil, society discombobulates, corruption grows, and governments fall. AI will ruthlessly accelerate this end to digital civilization as hostility grows and evil accelerates. Polarization emerges on every small and large issue, from international politics, religious edicts, and tribal culture to social norms. Increased conflict, tension, and hostility negatively affect social cohesion, law and order, good governance, and progress. It is the un-science of monolithic binary computers undermining stability, order, and tradition, driven by the darkest side of humanity. Under these conditions, AI society is self-destructive. The danger grows, and losses mount exponentially while democratic freedom, equality, and justice evaporate.

Powered by AI, everything about binary cyberspace worsens; there is only one way to prevent destruction. The science of nature is uniformly fair to every individual as a citizen of the world. Flawless computer science is neutral, guided by nature's natural laws. Mathematical science, when powered by the hands of private citizens, as taught in school, is flawlessly safe and needed to sustain a democracy. Then computer hardware physically guards the world's individuals against the dark sides of humanity using fair laws of mother nature. Precise science, undeniably equal to all as a fail-safe computer, offers privacy, security, and equality. Science is the only internationally acceptable, neutral solution to run trusted AI-enabled cyberspace.

With ever-increasing amounts of automation requiring sensitive data stored on computers, the rapid addition of AI increases corruption worldwide. Digital dictators enslave society but fail to protect our secrets. Keeping digital life safe and secure exceeds their ability. Life as a functioning, democratic cyber society depends on individuality and cyberspace's end-to-end stability, accuracy, and maintainability.

Eventually, AI will change every aspect of life, making democratic power vital to sustaining the most compelling yet dangerous machine ever built. It is globally impactful and thus worse than the atomic bomb. Civilization needs individuality, scientific equality, and freedom to implement democratic cybersociety through computational privacy and digital security. It is the only hope for worldwide peace and civilized progress. AI demands we rethink the cockpit of the digital computer to deliver open, accessible, safe, secure, civilized, international cyber societies.

Cybercriminals cannot have dangerous access to dictatorial powers within digital networks to steal sensitive information and silently use the unfair advantages of the superuser. It must stop; binary computers are flawed, digitally unsound, and democratically unsafe, while AI has the superhuman power and speed to run an Orwellian society. The rise of AI technology makes data collection and crimes effortless and instantaneous. Using deepfakes and pervasive cameras sabotaged to follow every movement, no one is safe at home, on the road, or at work; Big Brother exists and is constantly watching you. There is an understandable public outcry. Privacy is nonexistent, people's power is missing, digital security is nonexistent, binary computers are unsafe, and backward compatibility blocks constructive progress. Only the dark side of humanity wins.

The problem is computers are stuck in the past and made for crime, while AI exceeds human ability. While the software improved dramatically, and semiconductors shrank both cost and size, computers remain deliberately opaque, architecturally frozen by backward compatibility in the mainframe age of the 1960s. Binary computers are now outdated by AI. Updating hardware and software is costly and time-consuming. Furthermore, hardware changes disrupt captive markets. Soon, the effort needed to keep complex, obsolete hardware and monolithic software working will exceed society's capability. Skilled staff shortages will only grow. AI accelerates everything downhill toward national and international disasters.

Science is the only salvation, and the Church-Turing thesis, the cornerstone of computer science, holds the key to the stability of an AI-enabled, democratic society. Government regulation is needed to achieve democracy, or nations and civilization will end. The only hope is to rebuild computer science through the laws of science instead of ideology and commercial self-interests.

The story started in Babylon with the birth of the abacus.[16] It defines the first machine language that extended the power of individuals using arithmetic as computer science. This perfect machine of the age enabled civilization through international trade. Then, later, the functional machinery of the slide rule drove individuals to create the Industrial Revolution of the nineteenth century and the twentieth century, culminating with a race to the moon. Even functionally safe programming was understood and proved by Charles Babbage (1791–1871) and Ada Lovelace (1815–1852), two masterminds who cooperated on the mathematical machine code of a function-safe *Thinking machine.*

Artificial intelligence possesses an invisible evil like Alice's Red Queen, with additional superhuman powers of destruction that ethics cannot restrain. With instantaneous global reach to control digital society, corruption wins. Power corrupts, and the oppressive regime of unconstrained, superhuman AI let loose in binary cyberspace will crush civilization, bringing about Orwell's vision[17] of Big Brother through deep fakes, propaganda, and constant surveillance. Social media is full of disinformation. Denial of truth, doublethink, reinventing the past, creating the unperson, enforcing dictatorship, and ending democratic freedoms stop civilized progress.

Λ-calculus, the forgotten half of the Church-Turing thesis, is the scientific solution to all these problems. Defined as modular, protected digital abstractions, using the lambda symbol (λ), variables (a, b, c), and functions (f of x) to consume the variable. Furthermore, rules for reduction to the simplest form achieve significantly higher computation speeds. The λ-calculus frames the foundations of all mathematics. It enables the power of functional programming, improved programming languages, type theory, formal verification, and networking. These are essential to flawless computer science and secure cyberspace.

---

[16] Abacus, Wikipedia.

[17] *Orwellian* describes conditions George Orwell identified as destructive to a free and open society. It denotes a brutal policy of draconian control by propaganda, surveillance, disinformation, denial of truth (doublethink), and manipulation of the past, including the unperson—a person whose past atrocity is later idealized in public records, as practiced by repressive governments. It includes the circumstances he depicted in *1984*, but political doublespeak is criticized throughout his work.

The simplest example is the identity function. It returns its input unchanged, as in $\lambda x.x$, where x is variable. But what matters most to society is that each symbol is functionally secure and individually protected from corruption. Resistant to malware and fail-safe from programmed errors when engineered in hardware as a Church-Turing machine.

DEMOCRATIZATION

People Power

Computers and computer science did not begin with today's internet and the smartphone or binary microprocessors, which appeared in the 1970s. Or even in the 1950s when centralized mainframes ruled the batch-processing era and virtual memory begat privileged operating systems. Not with the Turing machine and the inventive genius of Alan Turing and his mentor, Alonzo Church, whose combined thesis in the 1930s defined computers as science and sparked the digital revolution.[18] Not even with Charles Babbage, the inventor of the *thinking machine,* a flawless mathematical engine he created at the zenith of the Industrial Revolution by the 1840s. Babbage's clockworks flawlessly abstracted addition, subtraction, multiplication, and division mechanically, and Ada Lovelace, Lord Byron's daughter nicknamed The Enchantress of Numbers, wrote and documented the first perfect, flawless scientific algorithm.[19]

Nor did John Napier, dubbed Marvelous Merchiston, the Scottish laird who, after two decades of arduous work, published his discovery of logarithms[20] in 1614. Or William Oughtred, the Anglican clergyman who used two logarithmic scales to create

---

[18] Church-Turing thesis.

[19] Ada Lovelace, the first computer programmer.

[20] His work, Mirifici Logarithmorum Canonis Descriptio (1614), contained fifty-seven pages of explanations and ninety pages of tables listing the natural logarithms of trigonometric functions he had calculated alone and by hand over two decades. Henry Briggs, the English mathematician, convinced Napier to change his logarithms to use the more common (base 10) logarithms. It took another two decades of hard work to achieve the same fourteen-digit accuracy.

the multifunctional, democratic slide rule[21] about a decade later. His flawless simplifications as the scientific slide rule accelerated engineering progress for three hundred years. From the Industrial Revolution of Victoriana and the two World Wars to the atomic bomb, the jet airplane, and the first moon landing[22] in 1969, the indispensable slide rule engineered progress.

Instead, computer science began several thousand years ago, long before these brilliant discoveries. Computer science started as a mechanical, functional language vital to the birth of civilization. To progress, Babylon democratized addition and subtraction as a robotic language of arithmetic and begat the dynamic framework of the abacus.

Language and trade combine distant communications with trusted computations. Thus, Babylon grew rich, and civilization evolved. The wheel automated transportation for those with roads, but the abacus sped and simplified arithmetic for all. The mystery of addition and subtraction abstracted as a trusted mechanical language proved easy to remember and use. Despite cultural differences, the trusted computation of large numbers enabled bulk trade through a common denominator—a human hand's abstraction. The abacus extended the power of individuals as independent, free-thinking citizens, each using a private thread of computation.

The hand abstracts just one mathematical type,[23] the decimal number. A fist for zero, increments through a thumb for five, to an open hand of nine, four fingers plus the thumb, the same way children first learn to count. However, with only two hands, the counting limit is ninety-nine, and traders demanded more.

---

[21] The slide rule is a mechanical computer for multiplication, division, and other functions like exponents, roots, and trigonometry. It is not for arithmetic that the abacus efficiently performs.

[22] In September 2007, Heritage Auction sold the Pickett slide rule that *Apollo 11* astronaut Edwin "Buzz" Aldrin took to the moon. The Pickett Model N600-ES, which was six inches long with twenty-two five-inch scales, then sold for $10.95. In 2007, at auction, it fetched $77,675.

[23] Type-safe computation is captured by R. Milne's statement that a well-typed program cannot "go wrong." Milner, Robin, "A Theory of Type Polymorphism in Programming," *Journal of Computer and System Sciences* 17, no. 3 (1978): 348–375.

Brilliantly, the abacus frames type-safe arithmetic for unlimited large numbers using a private chain or thread of mechanics perfected as the individual function abstractions, the type-safe decimal digits of any significant number. Functional perfection is the essence of flawless computer science. Later, the logarithmic scale enabled the type-safe functions of slide rules. Slide rules solve multiplication, division, and other perfect mathematical functions. Later, Ada Lovelace wrote not just the first program but a type-safe mathematical program using functions as variables.[24]

Sadly, this advanced capability, a century ahead of the science, an exclusive feature, and a distinguishing efficiency of the λ-calculus and functional programming, is still unavailable across cyberspace using binary computers. Church and Turing put all the pieces of computer science together scientifically by encapsulating the Turing machine within the λ-calculus. As previously exemplified when Ada Lovelace demonstrated functional programming, it remains unavailable to binary computers. They were frozen decades ago by selfish suppliers. Ada's algorithm, written in about 1840, passed scientific expressions as variables. Anonymous functions use symbols like the one Ada Lovelace defined for later resolution[25] once the values of n and B_1 are known.

$$f(x) = \{-1/2 *(2n-1)/(2n+1) + B_1 * (2n/2)\}$$

Without a centralized operating system and a compiler, all a binary computer can do is pass binary values within a single program. Functional programs and functional computers are scientific machines, far safer and more efficient than dangerous binary procedures. Binary procedures like Algebra only pass static values as variables, exposing binary values to whims, malware, and unfair digital privileges. Format differences permeate the branded un-science in binary computers and the loops within loops

---

[24] Functional programming is a programming paradigm constructed by composing mathematical and logical functions. It is a declarative mechanism using trees of expressions that map values to other values rather than a sequence of imperative statements updating the state of a binary computer that enables digital corruption.
[25] Taken from line eleven of Ada's twenty-five-line program to resolve the numbers of Bernoulli; see https://www.fourmilab.ch/babbage/figures/menat6_1-5k.png.

that impede performance and confuse individuals. For example, procedural programs resolve each variable at the earliest moment in a computational sequence. It significantly lowers the performance of step-by-step, binary procedures against the robust mathematical science of the λ-calculus.

The abacus began an unbroken chain of events that led to the λ-calculus and the Church-Turing thesis. Then things went wrong because von Neumann's overstretched shortcut ignored science. One must start with the abacus to understand the problems and appreciate the power of the λ-calculus and the Church-Turing thesis. And understand the natural implementation of science and the inherent advantages of private mathematical threads as abstractions using the scientific laws of nature.

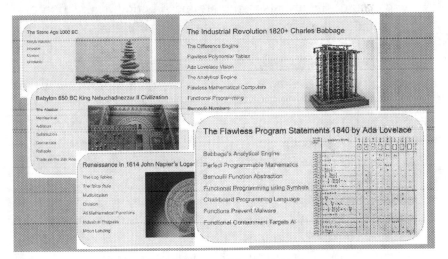

Figure 3. The History of Mechanical Computers

Inventive Babylonians elegantly designed an array of hands using as many rails as needed to count beyond ninety-nine. Importantly, each rail is a function abstraction of a single hand. So, the computational framework of a wooden abacus develops like silk threads, functionally woven and limitless. Consequently, this framework scales through the same atomic distribution and parallel concurrency defined by the λ-calculus and made available by the Church-Turing thesis.

The centralized framework of a binary computer only grows in opaque chunks, losing the advantages of nuclear scalability that dictate privacy and security. The atomic benefits lost in von Neumann's overstretched Turing machine cannot return because individual programming languages and stand-alone compiled binary images cannot support the power of networked λ-calculus. Monolithic, dictatorial operating systems block the way. After five decades of software progress that opened the door to artificial intelligence, the hardware remains trapped in the mainframe age before networking evolved and the full scope of computer science was unappreciated. The hardware industry still fails to employ functional programming and misses the native power of λ-calculus as legitimate, object-oriented machine code.

Each rail of the abacus is the atomic function abstraction of a human hand. It is a type-safe mechanical abstraction of a decimal number built according to the decimal type rules. Thus, each rail encapsulates one thumb bead and four finger beads and is limited to decimal numbers from zero to nine. However, far more critically, by starting with atomic scalability, every other aspect of computation is simplified, particularly privacy and security, while amplifying performance. The array of rails abstracts a limitless number, fueling trust in the accuracy of the more significant results needed by traders along the Silk Road.

The trusted abacus permanently replaced the stone piles in Babylon and worldwide. Stones compiled into haphazard heaps are mysterious, unstable, regularly fail, and invariably do not survive. One can only hope that sooner rather than later, the piles of untyped data in binary computers will also pass into history, replaced by the trusted mechanisms for software abstraction in Church-Turing machines. This change is vital for the promise of artificial intelligence without the downside risks and unintended consequences of an AI disaster[26] due to the insecurity of binary computers.

The faithful, functional laws of the abacus quickly built trust in working with large numbers, democratizing addition and subtraction, and expanding international trade. The trusted abacus became the

---

[26] An AI breakout is the loss of control over superhuman AI that has enormous unintended consequences.

engine of civilized progress, and computer science was born on the right foot.

Compiling stones into piles or, as today, compiling binary data into virtual machines is fragile and opaque. But pure computer science, as later defined by the work of Turing and Church, is simple and naturally distributed. The exact reasons the public accepted the type-safe computations of the abacus. It is a beautiful example of a scientific computer, thousands of years ahead of the theory established in 1936 as the Church-Turing thesis. The multipurpose slide rule again uses a type-safe set of private functions. Ultimately, implemented mechanically, computer science started with four core mathematics functions as Charles Babbage's *thinking machine*, explained and programmed by Ada Lovelace in the 1840s. Only these machines define computers scientifically.

Without transparent arithmetic, piling stones stalled market growth, and society was stuck with inadequate bartering. Such villages could not trade with the caravans and did not grow. Cities like Babylon needed bulk goods worth far more than two handfuls of decimal numbers. The camel trains needed the abacus to automate and mediate trusted prices for large purchases and sales between strangers speaking foreign languages. The abacus is internationally trusted because it democratized arithmetic. Easy to use, the abacus is graceful, faithful, reliable, atomic, and transparent. It is the scientific machine for decimal numbers and arithmetic that everyone learns the same way as a child.

Indeed, the user-friendly and trusted automation of type-safe mathematics will always accelerate civilized progress because simplicity democratizes computer science. Private threads as distributed computations equally empower every individual. On the other hand, the binary alternative is centralized, unfair, privileged, unscientific, crime-ridden, opaque, and overstretched. Binary computers only slow and stunt progress, leading society to criminal, industrial, and government dictatorships.

The public quickly understood that each rail of the abacus functions like a hand, abstracting and constraining the manipulation of a single digit as a number from zero to nine. Adding or subtracting large numbers is further simplified, reduced to sliding beads up and down rails, using the same carry and borrow rules learned by children at school.

Reducing complexity reduces the error rate and makes slips easy to spot and fix, one rail at a time, double-checked by the easy mathematical rules known to interested users. Each transparent step is an extension of the individual, a trader's private computational thread, adding or subtracting just two atomic numbers at any time. A threaded computation pioneered by the abacus is private and impervious to undetected interference and centralized catastrophes. The physical constraints are clear and unavoidable. It is impossible to bypass these physical laws because the end-to-end atomic structure enforces every computational step's scientific type and scope. Interference is detectable and preventable. The fundamental advantages of individual, cellular, type-safe calculations are physically private. They mimic fully protected λ-calculus threads, unlike the binary computer's centralized, unfair, shared computational architecture, exposed to outside attacks by malware.

Atomic mathematics avoids the distortions of the overstretching of the Turing machine. Atomic guard rails protect each instance of an abstract class. The abacus implements a decimal type. This mechanics of abstraction authorizes only three approved algorithms for a decimal number: add, subtract, and reset, guaranteeing no form of malware or AI-inspired interference. Each rail flawlessly performs addition or subtraction because foreign algorithms are not allowed. Indeed, engineered function abstractions only include approved algorithms. Thus, it is function-tight. There is no malware to worry about within the fully encapsulated types of a Church-Turing machine. Type-boundaries of object-oriented programs could prevent the threat of disruptions in a centralized binary compilation, but memory is dangerously shared instead. Furthermore, the computation of atomic types gracefully scales the capability, in this example, by another factor of ten without any algorithmic changes beyond the variable that defines the rail count.

While the abacus's storage mechanism of beads is physical, the algorithm is programmed, learned, and stored in a human brain, just like software programs in a computer. This transparent combination of individual atomic items bound together prevents outside interference, detects errors, simplifies checks, and empowers the flawless accuracy of basic arithmetic. From the earliest school days to university and post-graduate research, students learn the rules for perfect mathematics. Automating this same cellular process as a digital Church-Turing machine flawlessly extends the power of individuals instead of untrustworthy centralized organizations.

As proven by every historical success, atomic boundaries are vital for individual privacy, but digital boundaries are missing in the shared centralized compilation of a binary computer. Consequently, mistakes, spying, crimes, and cyber war lead to catastrophe. Such disasters will never end because the binary images are no better than a pile of stones in primitive Babylon's markets. The cost and effort to sustain these primitive binary computers are too high, and continuous patching and monthly upgrades will not fix the architectural shortfall. Binary computers do not represent actual computer science because they suffer from the undetected, unscientific disruption that empowers hackers and hidden digital crimes. The disturbance of digital data is undetected when attacked. Detecting every accidental and deliberate error demands atomic boundary walls scientifically defined by the λ-calculus to enforce type-safe functionality. Capability-based addressing[27] is the token-based hardware technology that realizes the goal.

Furthermore, enforcing type-safe boundaries does more than prevent outside interference. It guards the internal functionality of an encapsulated type while adding the advantages of functional programming to the machine code. In addition, a λ-calculus thread only places one digital data instance at risk in any computation. Every computation is a private thread with activity limited by need-to-know rules applied to each machine instruction. The dynamic assembly of the names in a hierarchical, nodal namespace implements this, the critical practice of top-secret security systems. Only authorized functions with approved access keys can reference a data instance.

---

[27] Capability-based addressing is a scheme used to control access to digital resources including memory. Pointers are replaced by protected, immutable tokens (called capabilities) created by secure abstractions in a Church-Turing machine, first demonstrated by the Plessey PP250. Instead of privileged instructions executed by a kernel or some privileged, superuser process in a computer. Tokens limit programs and subsystems access to approved digital resources and prevent write access when appropriate, without defining separate physical address spaces or performing a context switch.

Thus, approval exists before reading, writing, accessing, interpreting, or manipulating information in the type-safe cyberspace of Church-Turing machines.

When physical, type-safe boundaries double-check every atomic step, guaranteeing scientific functionality by detecting and preventing all sources of memory corruption, they achieve the same lightweight efficiency and effectiveness as the abacus. Capability-based addressing guarantees fail-safe, functional perfection built into the computer as a resident integrated development environment or IDE. Mechanized constraints function as in the abacus or the schoolroom teacher, detecting errors on the spot, blocking hackers and malware, and limiting mistakes to poor specifications and inadequate testing.

However, even when design bugs exist, they are immediately discovered by the built-in IDE as and when they occur. The flaw is pinpointed and quickly resolved individually, avoiding complex recompilations that expose other problems and demand extensive regression testing. Instead, retrieving individual digital objects from asynchronously updated home locations gracefully replaces outdated software and brings the utility of the cloud to all things in cyberspace.

Threading the atomic structure of a namespace as an individual, private, cellular computation avoids the centralized overheads, queuing delay, and shared boundaries invented for batch processing in the 1960s. Instead, the guardrails of atomic calculations prevent wild errors that pervade the outdated binary computer. Now, errors are limited, caught first, every time, by cross-checking every digital boundary. The technology of a typed computation further constrains each namespace through the need-to-know hierarchy of the application as individual calculations thread their independent paths through type-safe cyberspace.

Before the abacus, traders piled stones high to perform an addition above one hundred, but instability, confusion, exposure, and outside interference too quickly corrupted the compilations and the result. An accidental or deliberate kick in a dusty marketplace was all it took. Thus, the machine language of the abacus took off because the results were trustworthy, and the device remained the same over thousands of years, dependable and transparently easy to use. Together, the user and the abacus perform a thread of trusted computation. Even more interesting, as a type-safe computer, it follows mathematical laws later defined by the λ-calculus. Every calculation is atomically type-safe; thus, the architecture is far better than an exposed computer that runs without functional rules. Like the unstable stone pile in Babylon, shared binary compilations are equally unreliable. John von Neumann's computer, using the time-shared superuser for mainframes, froze out the progress of computer science over fifty years ago. At the same time, the abacus remained the trusted solution in remote village markets worldwide after thousands of years, billions of users, and endless correct results.

Indeed, every law of nature applies equally to all because the framework follows the cellular rules of mathematical function abstraction. Indeed, the computational framework of λ-calculus framed the abacus so long ago. It implements a thread of private calculations and avoids the unfair privilege modes that corrupt binary computers. These opaque, proprietary rules and unequal privileges ruin the foundation stone of digital democracy. Dictatorial centralization is unjust, proprietary, and biased. But the atomic mechanisms of mathematics are democratic, for one and all. Computer science lost its way and integrity when privileged operating systems became digital dictators, and the industry froze a faulty computer design. The abacus is profound precisely because users chose the intuitive wooden architecture to model the laws of the λ-calculus perfectly.

When Alan Turing's doctoral mentor, Alonzo Church, codified the λ-calculus at Princeton University in 1936, he led Alan to his Turing machine. Indeed, Alan's work meshed closely with Alonzo's. Alonzo's deep understanding of mathematics defined the computational model of computer science he called the λ-calculus. It drove Alan's binary model of algorithmic, threaded computations. Alan's simple Turing machine perfectly executes one function at a time, filling the need for the λ-calculus. The limited atomic nature of Alan's Turing machine is what Alonzo wanted and expected. The Turing machine is the λ in the λ-calculus, the private, secure, threaded engine of the Church-Turing thesis.

All this was below the horizon, technically too challenging, misunderstood, or forgotten when John von Neumann worked on the

ENIAC computer[28] at Pen's Moore School of Electrical Engineering. Instead of scaling through replication, distribution, and concurrency of private threads as ordained by the λ-calculus and demonstrated by the abacus, von Neumann dangerously overstretched the simplistic Turing machine. He created the centralized, vulnerable, shared computational architecture used today as a binary computer in many uniquely branded, dangerous, and confusing forms. The differences in all branded alternatives prove the un-science of ego-driven computer science today.

Thousands of years after the abacus, the cycle repeated when the brilliant, industrious Scotsman John Napier shared the secrets of logarithms in 1614. His formula turns multiplication and division into threaded steps of addition or subtraction. Born in Edinburgh, the eldest son of nobility, Napier became laird of Merchiston Castle, a pious reformer of God's universe. Long after his death, Napier's bones,[29] the computer framework he invented, remained superstitious magic to his grateful and respectful followers, including Samuel Pepys[30]. One hundred years later, Pierre-Simon Laplace (1749–1827) said, *"By shortening the labors, he doubled the life of the astronomer."*

However, Napier's bones, like a pile of stones or a compiled binary image, need specialized effort and peculiar skill to work reliably. William Oughtred[31] repackaged logarithms within a

---

[28] ENIAC (Electronic Numerical Integrator and Computer) was the first programmable, electronic, digital computer, completed in 1945. It was Turing complete and able to solve "a large class of numerical problems" through reprogramming, but it was Church-Turing incomplete that led to adding virtual memory and a privileged operating system.

[29] Napier's bones are a manually operated device created by John Napier for calculating products and quotients of numbers. The laborious method uses Arab mathematics and lattice multiplication.

[30] Samuel Pepys acquired a set of Napier's bones, as recorded in his diary entry of Thursday, September 26, 1667.

[31] William Oughtred (5 March 1574 – 30 June 1660), was an English mathematician and Anglican clergyman. After John Napier invented logarithms and Edmund Gunter created the logarithmic scales, Oughtred was the first to use two such scales sliding by one another to perform direct multiplication and division. He is credited with inventing the slide rule in about 1622. He also introduced the "×" symbol for multiplication and the abbreviations "sin" and "cos" for the sine and cosine functions. https://en.wikipedia.org/wiki/William_Oughtred

decade as the multifunctional, type-safe, simplified scales in a slide rule. He mechanized multiplication, division, square laws, cubed root extractions, sine, cosine, reciprocals, and more. His slide rule democratized mathematics for all. The slide rule performs any mathematical scale desired, democratizing the full range of mathematical functions for anyone to perform reliably.

Profoundly religious and fascinated by the Renaissance, Napier studied the night sky, searching for celestial relationships to confirm his belief in God's creations. He learned mathematics to understand astronomy. Moreover, he improved multiplication because the distances he calculated were huge. The movements of stars across the sky involve vast numbers, and time-consuming calculations made it difficult and slow, always performed by hand with paper, pen, and ink. The tedious work was error-prone, and the size of numbers stretched beyond the width of his parchment.

Napier questioned the method and thought freely. He simplified the slow and tricky procedures with an algorithm compressing numbers as an exponent of a base, then computed the result by adding or subtracting the exponents to find the result. Once the idea came to him, Napier could not rest. Working alone for over twenty years, he perfected the concept called logarithms, a word mashed together from Greek words for ratio and number.

By transforming to exponents, he raised the power of a base number to match a chosen value. In this novel form, multiplication and division simplify to adding or subtracting the exponent of the base power. For multiplication, he added lengths, as in 10^{23} multiplied by 10^{51} equals 10^{74}. Division of numbers uses subtraction. For example, 10^{23-51} equals 10^{-28}, where negative exponents represent a number less than one.

But he also went further, using properties in polynomial equations to extract square and cube roots quite mechanically. Every square root is the sum of two numbers, as schoolchildren learn from Pythagoras:[32]

$$(a + b)^2 = (a^2 + 2ab + b^2)$$

---

[32] Pythagoras: life, work, and achievement.

Napier's Bones allows users to guess one number and calculate the other. However, his clever bones still demand expertise, and just like the piles of stones in Babylon and the flaws in binary computers, the bones failed the tests of time. The type-safe slide rule embedded the algorithm as an unforgettable scale and universally replaced Napier's bones.

As a result, working with large, even enormous, astronomical numbers was simplified, and like arithmetic, the unskilled public could quickly calculate previously impossible results. When placed in charge of the Royal Navy, the diarist Samuel Pepys discovered the utility of Napier's logarithms and the *"Bones"* to estimate the volume of wood to build new ships. One only needed Napier's Bones or a logbook of tabular exponents following the logarithmic scale. Painstakingly, over two decades, Napier compiled this once and only once, then printed his book, *Rabdologia*, published in Latin in 1614. Napier changed mathematics in other ways, but his work on logarithms defines his genius. Other astronomers and scientists, including the English mathematician Henry Briggs, were at once captivated.

But Napier did not choose base ten. His choice of base 1/e was rooted in mathematical properties that offered advantages for certain types of calculations and table generation. While Napier's logarithms had technical advantages, it is less intuitive than the more commonly understood number ten. Henry Briggs recognized this and introduced this idea to Napier as more practical for day-to-day calculations due to the alignment with the popular decimal system. As a result, Napier ditched two decades of hard work and started again working with Henry Briggs to calculate the exponents for numbers to the base ten.

We face this problem again, choosing between imperfect binary computers and the scientific Church-Turing machine. Napier made his sacrifice in the best interests of collective progress. It is the same choice faced today to walk away from the shared binary computer for the altruistic improvement of cyber society and the good of humanity.

Briggs visited Napier twice before Napier's death in 1617, convincing Napier that base ten would be more familiar and thus practical. Napier agreed and abandoned his years of work to accept and work with Henry's idea. When Napier died, Briggs toiled on for seven more years, calculating over thirty thousand exponents to a base of ten by hand. Despite the limitations of that time, Briggs's logarithmic tables were a significant advancement and laid the foundation for more accurate and practical mathematical calculations.

Painstakingly, Briggs published the 500 pages of *Arithmetica Logarithmica* in 1624, computing, printing, and checking thousands of fourteen and fifteen-digit numbers printed as a logbook of pages. This endless set of strictly formatted numbers is another compiled static data set, opaque and peppered with hidden errors but, as ever, mindlessly trusted. Undetected errors lead to mistakes when computers crash. It may not happen at once but at an uncertain future time. The problem of hidden errors is the Babbage Conundrum, solved scientifically in 1840 by Charles Babbage's flawless *Thinking Machines*.

In 1622, eight years after Napier rocked mathematics, but before Briggs's logbook to base ten, another English mathematician, William Oughtred of Cambridge, simplified everything. He placed two logarithmic scales against each other to invent the slide rule. This device is another example of a mechanically type-safe computer. Better than the abacus and Napier's bones, the slide rule embeds the algorithm into sheltered computations, creating an even safer functional encapsulation of type-safe computing. Once again, using dynamic binding variables to the logarithmic scale removes all special skills needed to find results. The only thing to learn is the ability to slide two scales together using a cursor to select the variables.

Accurate results delivered instantly by the minor movement of the engineered slide is an elegantly uncomplicated design that democratized mathematics for all. It drove civilized prosperity to industrial levels for more than four centuries. Consistent and correct every time, without disconnected rods, stone piles, or complicated multistep procedures. One simple atomic action democratized more than multiplication and division but also square or cube roots of numbers and any other mathematical function.

The slide rule is the classic example of type-safe mathematical function abstraction scientifically codified by the λ-calculus. Artisans and engineers used Oughtred's slide rule for hundreds of years, carrying out remarkable feats of industrial engineering that, in

1969, at the birth of semiconductors, landed Buzz Aldrin on the moon. Until students backpacked electronics, they always carried a slide rule as the essential scientific aid that democratized computer science and mathematics worldwide.

THE RABBIT HOLE

The Path that Leads to Hell

Mechanical computers that began with the abacus and the slide rule ended after programmable computers began. Surprisingly, to some, this change started at the zenith of the Industrial Revolution with Charles Babbage's *thinking machines.* By the 1840s, Lord Byron's daughter Ada Lovelace proved the power of Babbage's mathematical machine code with her example program that calculates the numbers of the Bernoulli[33] series, documented as shown on page 43.[34] She noted the mathematics she used in Note E's "Statement of Results" column. She used a programmed loop from line thirteen to line twenty-three that (on line twenty-two) consumes the following functional expression as an unnamed variable:

$$anonymousvariable = \{A_0 + A_1B_1 + A_3B_3\}$$

The unnamed variable is a scientific expression, not a static binary value. It is more potent than the fixed values of binary computers. This dynamic function of named symbols is a complex operation. It remains unresolved, passed into the loop as a function abstraction for timely resolution later in the computation process.

This logical functionality as pure mathematical power disappeared when von Neumann redefined computer science as a stretched Turing machine, using shared memory instead of the λ-calculus. The analytical power of functional programs follows the

---

[33] The Bernoulli Numbers: A Brief Primer
[34] The first program, written and documented by Ada Lovelace.

symbolic laws of mathematics, while the physical alternative of a binary computer is limited to proprietary formats as specific values. Ergo, sharing creates an unsafe binary computer with physically different details, opaque complexity, limited insight, and problematic reprogramming. On the other hand, the abacus, the slide rule, and Ada's program last forever. The cost of ownership drops like a stone when computers function mathematically.

As students learn, mathematics and logic are flawless. Even artificial intelligence must obey the physical laws of science. Furthermore, engineers reliably built remarkable bridges and the tallest buildings engineered to function year after year for decades. Likewise, well-engineered scientific computers, including AI-enabled software, serve the public by preventing cybercrime. They correctly channel AI scientifically, preventing AI-breakout if computers were fail-safe Church-Turing machines. Other engineered devices work hard in factories, on roads, at sea, or in the air, and they work to meet every standard set for public safety.

Digital computers could do likewise. Malware is not a user failure, as claimed by experts. They shout, *"User error, they ignored best practices."* But infections contaminate brand-new computers before completing the first essential upgrade. Indeed, cyberattacks are nonstop, and computers that serve the public must cope with this, including AI-powered cyberattacks. The problem is there is no public standard for computer science, as for other industries serving the public.

Naturally, people assume computers work perfectly, but patched upgrades never end. Computer software, particularly AI, only passes the test of time when engineered scientifically to serve society with endless, fail-safe reliability. Using object-oriented machine code, the function-tight, data-tight digital objects have measured failure rates that exceed the underlying computer hardware to empower civilization. The advantages of AI, when safely engineered, will last forever. AI is a looming threat because the binary computer is flawed, and monolithic software reliability is the time between frequent upgrades. The technology of binary cyberspace allows AI to break out because it lacks detailed operational privacy and security.

When defined by the Church-Turing thesis, computer science bridges the gap between logic and physics, but binary computers ignore the logical side of this critical divide. Exclusively built on physical rules allows malware and crimes to escape missing logical limits. Cybercrimes are an ever-present danger, and AI will easily break out and escape civilized ethical boundaries.

Egocentric engineering and self-interested suppliers ignored the logical boundary problems in cyberspace to make the binary computer a rabbit hole to their branded digital prisons, run by dictatorial Mad Hatters from Lewis Carol's fables about childhood[35]. They celebrate daily the un-birthday of un-science. Corruption, to the point of catastrophe without warning, is amplified by AI. Alice's tutors become the Red Queen and Orwell's Big Brother, enabled by corrupt governments, industrialists, international criminal gangs, and foreign enemies. They are the only winners when the Mad Hatter is a dictatorial operating system. AI is the Red Queen of cyberspace, both good and bad, changing all the rules to gain selfish advantages while the games play out. Orwell's Big Brother is the dictatorial result of Cold War computer science.

Alonzo Church and Alan Turing formalized computer science as the Church-Turing thesis a decade before von Neumann hastily concocted the overstretched shared binary shortcut as the general-purpose computer.[36] Bridging the gap between logic and physics requires both sides of the Church-Turing thesis. But after World War II, binary computers began with a paper[37] by John von Neumann. He started again by ignoring science and overstretched the Turing machine. His result, the binary computer, forgot the symbolic, scientific half of mathematics and logic, defined by the λ-calculus in the Church-Turing thesis that hides the modular details through symbolic names.

When symbolic names hide digital memory, the binary implementation details become functionally secure assemblies of named black boxes. The binary computers expose these critical digital

[35] Lewis Carol, https://en.wikipedia.org/wiki/Lewis_Carroll

[36] Arthur Burks, J. Presper Eckert, Hermann Goldstine, and John Mauchly, along with numerous others, contributed to the creation of the first general-purpose, stored-program, electronic, digital computer. Von Neumann received the principal credit because he published the ideas to tell the world about them. He received credit because his reputation gave the greater weight to his words.

[37] Titled "Preliminary Discussion of the Logical Design of an Electronic Computing Instrument," by von Neumann, Burks and Goldstine.

details in shared physical memory, making it easy to accidentally or deliberately drop a digital spanner into the works. The shared memory architecture became a proprietary differentiator. Virtual memory has many branded variations created over the decades. It started with various sizes of memory word lengths, from bits to bytes to many bytes in length, including peculiar page-based virtual memory hardware mechanisms accessible to the superuser[38].

Virtual memory adds nothing to improve science. It only serves the industrial dictators who use them to lock in clients, sadly allowing the dark side of humanity to attack shared memory, including subverting the superuser with ransomware and botnets. The compiled image is no different from the opaque pile of stones in Babylon before the logical abacus. Nothing comprehensive prevents disruption or take-over. Binary machine instructions construct the physical address devoid of scientific, programmatic constraints, leaving hackers and criminals to ply their trade undetected. Undetected errors are the curse of binary computers and the un-science of cyberspace that leads to AI breakout, digital dictatorship, and national catastrophe.

Cybercrime exists because scientific digital guardrails are missing from binary computers. The breakout of AI is easy, specifically since AI already writes code of any kind. Blind trust allows hackers and their henchmen to disrupt these computers because they are all overstretched, shared, and centralized. These unsolved pitfalls only end in a dictatorship, AI breakout, or both. After half a century of building binary computers, the industry still fails to prevent cybercrime, including ransomware, and they never will. The centralized superuser is unstable. Repeated ransomware attacks prove this truth. The un-science of the centralized operating system and the binary computer will corrupt civilization beyond salvation.

Furthermore, the opaque, outdated software bugs require extraordinarily skilled human intervention, further complicating a tangled organization with severe staff shortages. Enhanced by AI will increase losses, need more urgent upgrades, and create unresolvable skill shortages. The unscientific playing field of the centralized binary

---

[38] https://www.quora.com/Can-some-one-explain-the-Von-Neumann-architecture.

computer tilts against innocent citizens to favor three increasingly dangerous groups: criminal gangs, industrial suppliers, and all forms of government. When enhanced by AI, only the citizens and democracy will suffer, and it will only get worse.

Privacy and freedom are needed in cyberspace to counter powerful dictators, but the superuser crushes both. Pervasive crimes overwhelm liberty, equality, and justice, killing the soul of democracy. Nations roll downhill as inscrutable dictators fight among themselves to find a winner, but without these democratic essentials, cyber society cannot progress.

Solving the security mystery for AI society requires equality. Cyberspace must be democratic. Equality exists in the bottom-up and top-down implementations of the Church-Turing thesis. The universally recognized work of Alan Turing ignited the digital age, but his mentor, Alonzo Church, is almost unknown. Church defined the science of a top-down architecture of secure software modularity throughout networked computer science. He called it the λ-calculus. The λ-calculus is pure computational logic, a mathematical science that precludes centralized, physical sharing by encapsulating functional software, good and bad. The trick of the λ-calculus enforced by Capability-based hardware and object-oriented software keeps the bad away from the good.

The computations are like the abacus and the slide rule: logical, private, individual computational threads. As learned at school, mathematical calculations are unique and personal, solved student by student, desktop by desktop, as private threads devoid of centralization and outside interference.

Moreover, again, as taught through mathematics, each scientific symbol stands alone as a black box, even for simple addition as used by the abacus:

$$a + b = c$$

Or for the area of a circle:

$$Area = \pi r^2$$

Or any other expressions of any functional complexity, for example:

$$f(x) = a_0 + \sum_{n=1}^{\infty} \left(a_n \cos \frac{n\pi x}{L} + b_n \sin \frac{n\pi x}{L} \right)$$

Each symbol is a functional black box solved in turn in a classroom and by the mathematical logic of the λ-calculus. No centralized sharing exists beyond rock-solid individual scientific functions that operate as private instances, but the scientific procedures never change, not in a lifetime, not ever. There is no pivotal shared or centralized point of binary failure, no language compiler beyond pure mathematics, no dictatorial operating system, the flawed essentials of binary computers, and their scientific downfall. The only open question is, will they take democracy down first?

By decentralizing software into Church's function abstractions, software expressions become digitally independent scientific functions. Functions link into expressions, as shown above, growing into application-oriented namespaces, like the global telecommunication application undertaken by the PP250[39]:

myCall = TelecommunicationNamespace.Call(+44-954-839-8555)

This powerful single-machine instruction is made readable and globally significant by the λ-calculus symbols of a functional telecommunications namespace. The unique names form hierarchical statements that keep detailed computations, including AI computations, between the guard rails of the application as a modular namespace. The pure, private, untarnished symbols keep calculations on track between scientific digital guardrails that consistently enforce the boundaries of the Church-Turing thesis.

Lord Byron's daughter, Ada Lovelace, wrote a symbolic, functional, mathematical program for Babbage's *thinking machine* at the height of the Industrial Revolution. She documented her 1842 program with the five detailed notes she added to her translation

---

[39] The first capability-based addressing computer, the PP250, was launched in 1972 by Plessey as a Fault-tolerant networked computer for Telecommunications switching across a network.

of Luigi Menabrea's *"Sketch of the Analytical Engine invented by Charles Babbage Esq.."*[40] This incredible document, including Ada's explanatory attachments A, B, C, D, and E, should be compulsory reading for every computer scientist. If von Neumann had appreciated her vision, the past century of lost progress would not have happened.

Ada programmed the *thinking machine* in twenty-five symbolic mathematical expressions without a centralized operating system or a language compiler. Babbage's scientific engine resolved terms privately, mechanically, and individually as a private thread. For example, she started with the scientific definition of Bernoulli numbers and used the same chalkboard mathematics understood by every scientist. Her program resolved each term to the ultimate result:

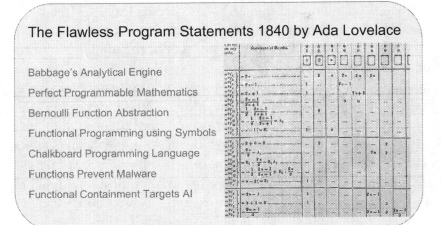

Figure 4. *Flawless Computer 1840 Ada Lovelace Machine Code of Chalkboard Mathematics*

The critical point is that the mechanics of scientific computers, like the abacus, are symbolic where:

$$(a + b) = c$$

---

[40] Ada Lovelace, Luigi Menabrea (1842), "Sketch of the Analytical Engine Invented by Charles Babbage Esq." *Scientific Memoirs*, Richard Taylor: 694.

The slide rule, where:

$$\log (m^*n) = \log (m) + \log (n)$$

And Ada's Bernoulli function for Babbage's unfinished *thinking machine* all work symbolically. In short, they all compute one symbol in one expression at a time and fully appreciate scientific equality as the = symbol. This solution removes all the dangerous, centralized baggage caused by hardware sharing in binary computers, including complex, shared virtual memory and opaque, dictatorial operating systems.

Centralization is the root cause of all the digital troubles in cyberspace. It is the curse of the binary computer. By ignoring the λ-calculus, Intel, AMD, ARM, and other microprocessors all work physically instead of logically. Thus, they all need a centralized, privileged operating system, such as Android, Unix, or Microsoft, before they can function as helpful, time-shared computers. The un-science of branded design is a product of the human ego. They only serve an engineer's or supplier's self-interest. After decades, familiarity and fear still lock clients to opaque computers that they struggle to understand, intentionally designed to limit progress that disrupts and undermines the existing computer business.

Alternatively, faithfully obeying the laws of mathematics as a machine removes the need for expertise in every obscure brand of binary computer and opaque operating system. Scientific function abstractions civilize and enhance society. Significantly, the first example, the abacus, democratized mathematics for everyone to use efficiently and accurately, accelerating trade along the Silk Road. Likewise, the slide rule hastened the Industrial Revolution and the space age. But, the binary computer froze the progress of computer science as a naïve, digital, cold-war mainframe. A first-generation design that discombobulates and corrupts cyberspace, deliberately and selfishly impeding progress. It turns everyday life into an ever-worsening Mad Hatter's tea party.

The lack of symbolic machinery confirms the un-science of each branded binary computer. Different binary formats and hardware details are brought to the fore, invented, and reinvented by every supplier in every decade. Each innovative design is as bad as others, caught in the dead-end canyon of von Neumann's binary shortcut.

Alan Turing proposed his binary computer to execute one algorithm in splendid isolation. His seminal work with Alonzo Church defined two complementary and mutually reinforcing essentials, a physical and a logical side of computer science. Without loss, a Church-Turing machine implements these two alternatives by encapsulating a simple Turing machine as a λ engine in the λ-calculus. Their combined work hides the dangers of implementation within the logic of comprehensible scientific notation as modular function abstractions. The same method learned as a child. The clockworks of each λ-calculus expression automatically time-share the most uncomplicated, least-cost Turing machine as a simple engine of pure, private computational threads. Everything else results from the remarkable powers of the λ-calculus and immutable capability tokens.

The Turing machine is the bottom-up design, physically connected to the natural world using hardware registers and device drivers. But as an encapsulated, one-program-at-a-time engine, it becomes the λ-calculus of the λ-calculus. Program binding, defined by a namespace instead of by compiled code, acts as a road map for cyberspace. Each mathematical name forms a fail-safe expression as an algorithm that stands alone and runs in isolation as the λ step in a cellular thread of computation like the mechanical cells of the abacus.

The λ-calculus is the top-down mathematical science of the Church-Turing thesis defined by Alonzo Church. He expressed all the scientific rules for distributed, modular, asynchronous, parallel computations in a functional namespace of dynamic threads and, thus, perfectly hides Turing's binary computer. Everything needed for flawless computer science is democratic, requiring no superuser. Better than centralization, the λ-calculus directly supports functionally distributed programming[41] performed in machine code through programmed object-oriented abstractions. When copied and networked as individually threaded communicating computations, the same asynchronous, distributed computational threads

---

[41] Functional programming is a programming paradigm where programs are constructed logically by composing functions. It is a declarative programming paradigm in which function definitions are trees of expressions in a namespace that maps named objects to other named objects, rather than a sequence of imperative statements that update the running physical state of the program, as used by binary computers.

performed in Babylon and schoolrooms exist. Each calculation is computed privately in a protected chain of crime-free, fail-safe, data-tight abstractions. Threads weave their path like silk throughout the namespace and beyond to other black-box objects anywhere in known cyberspace.

Consequently, all the complicated and dangerous branded baggage repeatedly reinvented by every supplier is a dead-end canyon, just a flawed shortcut like some pioneer trail, littered with the discarded junk of misguided progress. The binary computer is just a branded Mad Hatter,[42] scientifically no better than the piles of stones once used for counting in Babylon before the functional, everlasting, mathematical abacus.

Princeton University is home to brilliant ideas in conjunction with significant controversy between the egotistical staff members who all see things differently. The feud between Robert Oppenheimer and Lewis Strauss is one example[43]. Like atomic science, computer science would change the world, and John von Neumann's ego wanted his name to be first, as was Oppenheimer's after the Atomic Bomb that ended World War II. For this, John von Neumann saw no value in his colleague Alonzo Church's work, specifically the need for the scientific guardrails of the λ-calculus. In the race to be famous, von Neumann removed Alonzo Church's guard rails, vital for industrial-strength computer science. His shortcut misled the world to build the first-generation binary computers simply by overstretching the Turing machine.

Now hackers and malware freely roam cyberspace, like the highwaymen of old. Omitting the λ-calculus overstretched and consequently overshared the simple Turing machine designed for a single algorithm. Over the decades, binary programs grew into massive, mysterious virtual machines. Throughout time, the imperative machine code remains unguarded and dangerously

---

[42] Until the twentieth century, hat makers used mercury to stiffen felt hats. They used mercuric nitrate in poorly ventilated rooms. Over time, the hatters inhaled mercury vapors and developed chronic mercury poisoning, including psychosis, excitability, and tremors, so common the phrase "mad as a hatter" was born.
[43] How Lewis Strauss Shaped the Atomic Age—and Orchestrated Robert Oppenheimer's Downfall https://www.history.com/news/lewis-strauss-nuclear-program-oppenheimer

misused by programs in every binary computer simply because the symbolic addressing mechanisms of mathematics and the λ-calculus are missing. The absence of mathematical symbols, measured by malware, hacking, and cybercrime, reveals the insecurity of centralized binary computers. It is already a costly disaster, becoming an unmanageable catastrophe when AI runs the world.

Without digital guardrails, the software is unchecked. Blind trust in high-quality best practices is an insufficient hope for a world of AI-enabled cybercrimes and international conflict. The ingenuity of encapsulation employed by the Church-Turing thesis deploys Alonzo's top-down functional modularity with Alan's bottom-up machinery. As, where, and when science matters most, on the hard edge of a digital computer as programs execute and digital power is released, the laws of nature apply in full. Changing the machine code from opaque, imperative binary commands to declarative statements is fundamental because declarative statements are targeted to scientific symbols by name instead of a commonly shared binary compilation. When software and hardware interact symbolically on the hard edge of computer science, double-checking everything takes place before the computer brings a program to life. The binary computer lacks this built-in security. Using binary machine code is too challenging to understand, opaque to program, and dangerous for public service. It's as unpredictable as the Mad Hatter in Lewis Carroll's topsy-turvy Wonderland.

The computer industry fell down this rabbit hole dug by von Neumann, followed by universities, businesses, individuals, and nations expecting experts to have figured out a scientifically robust solution. It failed to happen. The networked Turing machine is a Swiss cheese full of voids quickly filled by hackers and crooks, creating bizarre and irrational software experiences. Vulnerable programs, program mistakes, bugs, and deliberately crafted attacks collude, corrupt, steal, spy, and silently perform unspeakable, undetected cybercrimes.

Binary computers use imperative instructions with direct physical access to shared virtual memory, into which compilers pack programs cheek to jowl as a binary image for the binary computer. Program A and malware B share access to everything in the virtual machine, including all other data and programs. Undetected malware uses these shared physical powers to attack the virtual machine.

The physical instructions include mathematical and logical commands and other proprietary control commands to supervise program steps and control the branded hardware and any attached physical equipment. All these binary instructions share the overlapping default powers of shared physical addresses, easily forged by mistake or design. However, the peculiar, branded machine instructions to control the superuser and virtual memory are so dangerous that only a trusted operating system should use them. These commands hide as a privileged superuser, but these proprietary mechanisms cannot avoid the so-called zero-day attacks.

For example, the hidden mechanisms of the operating system perform time sharing between virtual machines, change virtual memory maps, and regulate the power of the enslaved virtual machines. To carry out these hidden services, users must switch to the operating system and run as the superuser, switching virtual machines and changing virtual memory. The remaining RISC[44] instructions are programmer accessible, but they all share the default, mindlessly shared powers of the physical computer, including immediate unchecked access to the memory of other programs. Enemy malware attacks these default powers, including other inherited user or superuser powers.

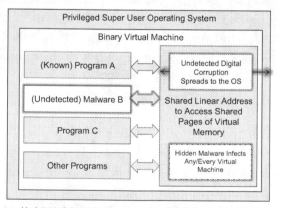

Undetected corruption from as small as a bit to the size of the
Virtual Machine's memory space includes undetected malware that
causes ever more problems, spreading cybercrime, computer
crashes, and AI breakout through the OS

Figure 5. The shared memory problem in binary computers.

---

[44] "What is a Reduced Instruction Set Computer (RISC)?" Definition from Techopedia.

In the early days, program size often exceeded the availability of computer memory, leading to complex virtual memory hardware organized like the pages of a large book, with most pages unopened, kept on backing storage. The central operating system manages all the extra dangerous hardware as a superuser. All this complex physical machinery is proprietary. It adds no value to computer science, as defined logically by the λ-calculus. When a single point of failure breaks down, ransomware freezes everything, and AI-enabled malware will make this attack more frequent and even harder to avoid.

When a binary computer addresses a missing page, the operating system opens the page from remote storage. It adjusts the memory maps before returning smoothly to the interrupted virtual machine. The centralized operating system running as a superuser controls this hardware paging mechanism, with the power of the privileged instructions to turn the pages on request. The superuser hides this complexity from user programs. Only superusers should access the branded hardware mechanisms invented by each supplier, but these dangerous mechanics are exposed when malware penetrates a superuser and ransomware, or worse, results.

Without the centralized operating system and the branded powers of the hidden time-shared services, binary computers remain the simple Turing machine. A small, stand-alone, single algorithm on a paper-tape solution, as conceived by Alan Turing for his mentor to automate the λ-calculus. Without the dynamic ability of the λ-calculus, programmers wrote more extensive monolithic procedures to move the software between local and remote memory units. Thus, a centralized operating system simplified this work by standardizing many complex tasks and reducing effort. Soon, centralized monitors checked requests against approved lists to confirm a user's access rights to the centralized equipment and shared services, adding another overhead that poorly replaces the missing modularity of λ-calculus symbols. Indeed, the gravitational pull of functionality towards the superuser is a black hole in cyberspace from which, eventually, nothing escapes.

Today, branded operating systems define the black holes of cyberspace, growing ever more complex and unsatisfactory as functions enter their orbit. They perform an increasing set of shared complex services that demand superuser status. In the beginning, they only replaced human operators. First, an administrator to log in users; second, a technician installs and runs applications as virtual machines. Finally, as a magician, time-sharing the computer by swapping virtual memory pages between virtual machines. Centralized work also includes responding to users' keyboard and mouse requests, managing communication ports and printers, and creating images on computer screens. Indeed, operating systems now perform every important, common task on behalf of all users in real-time, as and when needed. Still, experience shows they fail at user privacy, detecting malware, and computer security.

These centralization benefits came at a heavy price that only worsens with AI-enabled cybercrime. The central operating system is an unstable overhead and a single point of failure. It must change whenever device driver services change or a new criminal attack succeeds. The centralized operating system is as fickle as the uncertain stone piles in Babylon.

This problem led suppliers to backward compatibility, limiting changes, freezing hardware progress, and still running old programs. Suppliers learned that backward compatibility protects their market by locking in their client base and restricting changes to ones they want, not ones that benefit others. Worse still, when a design flaw exists, the operating system is attacked and replaced by a criminal or enemy alternative. This ultimate assault is a coup d'état,[45] implemented as part of a ransomware attack or to turn the binary computer into a zombie enslaved to some international criminal gang or enemy government for spying and spam.

Users should not run as superusers, but the choice is not easy. The competing demands are complex, and the obscure skills needed become more opaque and harder to apply. Running as a superuser is easy because security settings do not apply. Life in cyberspace is easy but far more dangerous. When users run as a superuser, the malware threat reaches every corner of the binary computer. It stretches to remote memory and all attached equipment connected

---

[45] Coup d'état, the sudden seizure of power that redefines the binary computer as a zombie controlled by criminals over the network from afar.

worldwide, as proven by the Stuxnet[46] attack on Iran's uranium centrifuges. For individuals and corporations, ransomware attacks usurp the operating system and then scramble the binary data on the backing store, halting all operations until the specific demands of an unknown attacker are met.

Thus, significant overheads exist for the centralized operating system to check every request and mediate interactions laboriously. In this process, any attempt that exceeds the authority of the user identity is blocked, including communications with a networked endpoint using one of many different protocols. However, mediation cannot check if malware is already present. Without the overhead of conciliation, a binary computer has no digital security. Even more unfortunate, branded protection is limited to the gross needs of all programs instead of a particular program's atomic requirement. Even with an excellent operating system and the best networking security systems, AI malware will still attack all the resources available to the user.

Digital security cannot exist when any program that takes control shares the authority of a virtual machine through the physical addressing space. Thus, attacks start quickly in any virtual machine contaminated by malware by misusing the imperative binary commands and the shared virtual machine invented by each branded supplier.

Instead, meaningful protection must apply to the individual needs of logical software modules. It is the essence of λ-calculus symbols and expressions. To detect and prevent AI malware, identifying the details and following the laws of the λ-calculus is vital.

Pervasive digital security is modular and symbolic. It must begin at the edge of the digital computer, where hardware and software touch, where every digital mistake originates, and digital corruption starts. It is at this hard edge of computer science where the Church-Turing thesis is needed most. If AI-enabled cybercrime and AI cyberwars attack flawed computer science, the end of cyber society will come fast. Enemy states that sponsor terrorism and war will

---

[46] Stuxnet, Wikipedia: Stuxnet is a malicious computer worm uncovered in 2010 and thought to have been in development since at least 2005. Stuxnet targets supervisory control and data acquisition (SCADA) systems and is believed to be responsible for causing substantial damage to the nuclear program of Iran.

turn to massive cyber-attacks powered by AI to get their way, and the only scientific option is to add the guardrails of the λ-calculus. Cyber threats are detected when the machine instructions empower the programmer to protect their code from attack. Enforced modularity includes enforced functionality of λ-calculus as fail-safe units of digital security for cyber-democracy to survive and flourish in peace.

Furthermore, the symbols from the λ-calculus now exist in machine code, making machine code readable by anyone interested. Each name, as a symbol, defines a typed security boundary for a named digital object. It is how the Church-Turing thesis democratizes complexity and simplifies computer science. It is the same successful strategy achieved throughout the millennia of mechanical, clockwork computers that began in Babylon. Three generations of simple, scientific solutions started with the abacus and later progressed through the algorithmic slide rule to culminate with Babbage's programmable *thinking machine*. Democratizing cyberspace this way demands that the digital computer employ the mechanics of λ-calculus. If so, even amateurs will spontaneously write functionally fail-safe programs. Automatic functional security accelerates the training and prosperity of civilized cyber society in unimaginable ways, with unbelievable advantages.

As programmable software evolves and cyberspace expands, every proprietary computer has a changing set of flaws. Complex flaws are discovered first by criminals. Manufacturers, government spies, hackers, warriors, and enemies, all enhanced by AI, use these defects to achieve a selfish advantage. Consider identity-based virtual machines. Security is limited to the identity plus a password, where the identity defines the approved services centrally described for each virtual machine. But this is insufficient. It is the actions of individual programs where all the trouble starts. Atomic identity is precisely how the distributed, concurrent advantages of λ-calculus security work. Instead, identity-based access management[47] (IAM) is centralized and once again supervised, administered, and checked as lists of approved users by a disinterested digital operating system. The theft of these centralized credentials in a data breach has a long-

---

[47] https://www.microsoft.com/en-us/security/business/security-101/what-is-identity-access-management-iam

term impact because credential exploits echo through cyberspace long after the robbery.

But credentials are not vital to exploit an account because back doors are left wide open. Trouble starts because networked access to the imperative machine instructions is mindlessly trusted. Any code downloaded from an email when browsing the internet or by file sharing automatically inherits the user's identity and privileges plus the default power of the physical machine. Even a trivial desktop game like Minecraft has the authority to use digital resources and rights to access private files and discover digital secrets. In this interconnected, digitally converged world of computer science, conflict exists because hackers, enemies, and criminals share the physical computer but do not follow best practices. Instead, the malware uses any default power to attack the computer, attached equipment, the operating system, the user data, and cyberspace in any crazy, unexpected way.

Designed when hardware was limited and expensive and software was free, the tables have turned, but this advantage remains ignored by the backward design. Developing monolithic software is costly and cumbersome, while semiconductor hardware is almost free, switching the balance of power to using the atomic modularity of the λ-calculus. But the irrationally shared binary computer remains frozen in place. The outdated architecture from the Cold War is a looming problem for AI society and the progress of democracy. The operating system is blind to the digital voids, open back doors, and inherited resources in the software overstretched across cyberspace.

Every operating system must interwork with all other brands and every network protocol used to manage access and distribute files and services. Every computation shares a user's authority and can do whatever they choose within these ever-expanding gross limits. The patchwork approach to cyberspace and compiled machines is unavoidably flawed. Tolerating undetected digital interference within a user's private space is only the beginning. When overpowered by AI, centralized patchwork powers suffocate progress and destroy democracy.

Blind spots pervade patchworked cyberspace. In each case, essential information is missing when the rubber hits the road and the superuser mishandles a request. This systemic problem is called the confused deputy attack, [48] [49] caused by the patchwork boundaries created by suppliers. Malware dynamically undermines these static privileges on the fly by directing requests to the operating system through a third party that escalates the demand to a superuser.

It is a systemic problem with Identity-based security. Cybercrimes exploit this as a clickjacking[50] attack or cross-site request forgery[51] attack in conjunction with phishing[52] attacks. The tilted playing field favors governments, suppliers, dictators, international criminals, spies, and enemies, but not innocent citizens. Suppliers cling to the past to slow the progress of computer science while individuals pay the price of lost freedom and missing privacy. At the same time, society bears the additional cost of dictatorial domination by industry, criminals, and foreign enemies. Competition and progress will grind to a debilitating halt as creativity is overwhelmed by endless maintenance from increasing AI-enabled attacks. A report by Accenture estimated that the average cost of cybercrime for organizations in 2020 was $13 million. It includes the direct costs

---

[48] The confused deputy attack—a cross-site request forgery (CSRF) is an example of a confused deputy attack that uses the web browser to perform sensitive actions against a web application. A common form of this attack occurs when a web application uses a cookie to authenticate all requests transmitted by a browser.

[49] https://en.wikipedia.org/wiki/Confused_deputy_problem

[50] Clickjacking is when harmless features of webpages perform unexpected actions to trick a user into an undesired action hidden in a concealed link. On a clickjacked page, the attack loads a transparent page over the visible page. They assume they click the visible buttons, but they are actually activating the transparent page. If the hidden page is over an authentic, the attacker tricks users to perform actions never intended.

[51] Cross-site request forgery, also known as one-click attacks (CSRF), is a malicious website exploit where unauthorized commands are transmitted from a user. Unlike cross-site scripting (XSS), which exploits the trust users have for a particular site, while CSRF exploits trust a site has in particular browsers. About eighteen million users of eBay lost personal information in February 2008. Customers of a bank in Mexico were attacked in early 2008 with an image tag in an email. The link in the image tag changed the DNS entry for the bank in a router to point to a malicious website impersonating the bank.

[52] Phishing attacks capture secrets like usernames, passwords, and credit card details by masquerading as trustworthy sources, luring unsuspecting targets to infected websites that download malware. Email spoofing and instant messaging are examples that trick users to enter secrets at fake websites that look and feel like the legitimate site.

of responding to and remediating a cyberattack and the indirect costs associated with lost productivity, reputation damage, and other factors.

While computer hardware remains stuck in the days of the Cold War mainframe, software progress has accelerated and will do so again. AI is the next step in the endless chain of potential improvements to software, introducing new threats. According to a report by Cybersecurity Ventures, the global cost of cybercrime will reach $10.5 trillion annually by 2025. A fraction of this cost will pay for the innovative technology required to fix the problem at the machine level. At the same time, human support costs will fall dramatically. The only acceptable solution is science—engineered digital security defined by λ-calculus and following the Church-Turing thesis. By 2025, humanity's collective data will reach 175 zettabytes—the number 175 followed by twenty-one zeros; this includes everything from government data, streaming videos, and dating apps to healthcare databases. Securing all this data is vital to a functioning cyber society. Meanwhile, insecurity continues to grow.

THE RED QUEEN

The Rules Keep Changing

The promise of cyberspace as an extension of individuality cannot prosper when enslaved to industrial dictators. Flawless, democratic cyberspace is urgent and vital because dependent users and computerized society are only extensions of industrial dictatorships. Enslaved by favorite apps and spied on by criminals, dictators are the winners with a cabal of criminal henchmen. In a reversal of fate, users become tools, extending their reach and motives (if not the substance of cyberspace) back into the physical world. Dramatically proven by Stuxnet in 2005, the power of software extends already in the physical world. The whims of an AI-empowered Red Queen[53] now overdrive mortals and machines from within cyberspace.

Life becomes *real virtuality*, the reverse of virtual reality. Not reality as it was but changed by the Red Queen. Apps gifted by dictators redirect users to follow the direction of the branded dictators. The Red Queen appears from Alice's mirror instead of Alice entering her looking glass.

In this reversal, individuals are business assets living in the natural world but acting as the agents of a dictator. We turn left when the GPS mounted in the car tells us to do so. We buy gas and coffee, following the instructions of a favored app. We bank, eat, fly, and interact socially according to the virtual reality we use—software

---

[53] Lewis Carroll's book *Through the Looking-Glass* focuses on a child's loss of innocence, the rules that govern the world, and the struggle to follow rules that are unexplained. When Alice enters the looking-glass, she enters the strange world of the Red Queen, where nothing is what it seems, and everything is distorted.

that, in the main, is not written in our individual best interests but for another purpose. To push the sale of things not wanted, to give up what is private and personal, and to surrender to the ideas and wishes of the dictatorial developer.

Are deaths in school shootings caused by atrocious games played in the metaverse? Do democratic governments spy in secret on their innocent citizens? How many votes are manipulated by social media? Marshal McLuhan told us in his 1964 seminal book, *Understanding Media: The Extensions of Man*, that *"the medium is the message."*[54] If corruption and dictatorship remain the message of cyberspace, civilization is doomed.

Virtual reality was once an art form that imitated life, but now life enacts the reflections of cyberspace. Everyone responds to cyber applications, from global finance to local utilities and central government to daily shopping, family health care, and gambling with strangers. As we lust for physical and emotional satisfaction, we increasingly use cyberspace. We are dependent to the degree that it is dangerous for the survival of our shared natural, national, and personal interests. The danger is not just because we are addicted to computer science but because trusted software does not exist in a binary computer. Yet, we mindlessly follow as prisoners of the Red Queen. Digital integrity is lacking because we share memory space with dangerous programs pushing uncertain motives, all run by corrupt digital dictators.

We depend upon this unsafe binary concoction in every occupation. With social media, in one extreme form, individuals, either too young to know better or adults too vain to be humble, send sexually explicit messages, including photographs, dubbed as *sexting*. Gibson's vision is right[55], and we have indeed plugged into cyberspace if happiness (either sexual or emotional) includes sexting. Life-supporting services from food to water, from police to fire brigades, and from governments to national defense are all computer-dependent. They depend on unreliable digital services drawn from

---

[54] The medium is the message, Wikipedia.
[55] The term cyberspace first used by the American-Canadian author William Gibson in 1982 in a story published in Omni magazine and then in his book Neuromancer. In this science-fiction novel, Gibson described cyberspace as the creation of a computer network in a world filled with artificially intelligent beings.

dictatorially corrupt digital cyberspace. In all conceivable ways, a Red Queen decides what happens, and when the computer is down, life comes to a screeching and unsettling halt.

The Red Queen is the spirit of software corruption with disturbing peculiarities. Dramatic and unexpected service failure is one. Undetected errors and undetected digital interference are other examples. Distorted human behavior like sexting is another. There are no natural alternatives to these alarming peculiarities when depending on cyberspace. Abandonment of standards is the original cause. It is a consequence of unsafe engineering pervading the workings of cyberspace. It emanates from corruption to impact humanity into dangerous, unnatural behavior.

Sexting, as a word in the modern lexicon, is like the words for logarithm and mash-up. These words speak their idea from two other words, sex and texting in one case, or for Napier, who defined logarithms from the Greek logos and arithmos.[56] Texting is also a mash-up, peculiar to cyberspace. It is what one does, or at least what our children do with thumbs and fingers, when using a smartphone while walking, talking, driving, and socializing. Texting while driving proves that best practices and cyberspace do not work together. As Gibson foresaw, cyberspace changes human existence in strange ways. We are a computer-dependent society because virtual reality is easily absorbed without deep consideration. Driven by a Red Queen, life becomes an extension of the dictator's mash-up, simultaneously with incredible successes and catastrophic failures.

Thwarting the Red Queen is vital for human development and for advancing civilization. It depends on decentralization and the replacement of dictators with democracy. Democratic societies must survive. Specifically, freedom and equality must reign, and cyber dictators must vanish, replaced by people power. The science of Alonzo Church and Alan Turing allows both goals at once. Digital technology can build the best practices of computer science into the cockpits of each digital computer, reproducing Babbage's flawless computer science by detecting corruption on the spot. Cross-checking the two sides of a Church-Turing machine against each other as actions occur in the cockpit guarantees attacks cannot spoof

[56] Napier was, as discussed elsewhere, was well ahead of his time in so many ways.

a user because perfect computer science is, by definition, flawless. Digital integrity is essential, nay vital. Otherwise, corrupt cyberspace is dangerous, hazardous to humans, and unacceptable as a dictator of individual, social, national, and international behavior.

The word *mash-up* originated in the music industry. It describes harmony created by mixing soundtracks played both forward and backward, then blended with lights to the beat of the music, popularized by DJs in clubs during the 1980s. It migrated to the internet to describe a service that, as unexpectedly as sexting, combines ideas, meaning, data, presentation, sounds, and functionality drawn on demand across cyberspace. How different is this from John Glenn's[57] era when isolated mainframes and batch-processing computers were limited, like the cramped quarters of Mercury-6?

Nothing is certain in this new world order because everything is dynamic and on demand. Nothing can be planned and pretested or trusted. Mash-up captures the spirit of the internet, searching and browsing, and rapid evolution that exists when exploiting digital convergence and the World Wide Web to the limit. Mash-up is a label for changing times that redefined computer science from bits and pieces like computers and phones. These objects of yesterday's world have morphed into an abstract galaxy of user interfaces, digital information, services, and programs, good and evil, trusted and dangerous, driving all shades of life with motives populated by global cyberspace. It is a profound change in thinking brought about by international digital convergence.

Almost a century ago, the digital generation took off with Alan Turing. A decade later, after World War II, John von Neumann defined the binary computer and compilers, and programming languages began. In the 1950s, time-sharing and batch-processing computers began the mainframe era. The refined form of mainframes created minicomputers, then semiconductors configured as virtual machines with virtual memory. The Red Queen appeared through the magic of high-density semiconductors, networks, and the privileges of dictatorial operating systems. None of this garbage was needed when Ada Lovelace wrote the first functional program in the previous

---

[57] Mercury Atlas 6 (MA-6, also designated Friendship 7) was the first orbital flight of an American rocket with a human on board. The pilot was John H. Glenn, Jr.

century; she used what Babbage provided as a machine language, the scientific symbols of mathematics and logic.

The illusion of the dictatorial operating system regulates space and time for several user tasks to run simultaneously. This dangerously alternative digital world created on the fly by the human ego designed the operating system to rule the binary users of virtual machines. But the logged-on user of today is no longer alone in cyberspace, as in the Cold War days of John Glenn and von Neumann. Today, we share a binary computer with strange programs of unknown providence that avoid best practices to corrupt results. The trip gets ever more dangerous as shared digital space is united as a user desktop, within corporations, and worldwide in overlapping social groups. The related computations share the full authority of the user with access to every desktop asset and, all too often, elevated superuser powers that cut into far larger digital voids that span cyberspace.

As continuously reconfirmed by growing cybercrime, the binary computer does not make a passing grade as the cyber-ship for the digital mash-up. The binary computer exposes every user to malware added to the desktop by browsing, emailing, texting, and sexting. The desktop is invaded every day by untrustworthy code. Malware moves in to stay because it can easily use the dictatorial powers of the Red Queen to evade detection. AI malware will go further. Binary computers cannot enforce an execution model matching the mash-up; worse, users do not respect best practices when absorbed by an addictive way of life.

Cyberspace is an unnatural human experience, a foreign place far changed by networking from when Multics[58] redefined binary computers for batch processing mainframes. These time-shared mainframes are out of their depth in networked cyberspace. A cockpit revolution is needed for humans to journey through global cyberspace safely. To safely experience the galaxy of good, bad, and evil information that stretches into the endless future, data-tight,

---

[58] Multics (Multiplexed Information and Computing Service) was the comprehensive, general-purpose programming system developed as a research project by MIT. The initial Multics system targeted GE 645 mainframe computer. The Multics operating system implemented multilevel security (MLS) on a GE 635 by running a simulator of the 645 starting on October 18, 1965, in the MIT Tech Center. None of this work adequately considered networked malware.

function-tight, fail-safe Church-Turing machines are crucial. The type-safe architecture of engineered checks and balances inspired by the Church-Turing thesis is for the dangerous, unknown digital journeys on which every individual and collectively every nation has embarked.

Mercury-6 was airtight to prevent threats on the first human trips into outer space. Likewise, travelers in cyberspace require an environmentally engineered computer as a fail-safe machine to avoid the digital threats of cyberspace. For all future time, the cockpit of a digital computer and every machine instruction must be data-tight, function-tight, type-safe, and fail-safe. Only then will malware be recognized as program errors identified, detected, and prevented immediately. Civilizing the Red Queen needs a level field for private individual computations that obey the foundational work of Alonzo Church and Alan Turing.

The security and freedom of the unskilled public as a democratic society when using dangerous software as a public utility is an enormous unresolved national concern only resolved by science and the networking power of the λ-calculus. On-demand networked transactions, continuously delivered by the mash-up, must be safe for public use, accessible, and efficient for any unskilled citizen or amateur programmer. The Church-Turing machine powers individuals privately without any downside, from a child's first lessons to experiments and research in any field of knowledge. Best practices are enforced automatically as an engineered science. The industry is responsible for a trusted design only achieved when verification is automatic. The government is responsible for setting the scientific standards of cyberspace as a Public Utility for endless democratic civilization. It is vital for the survival of democracy, a national concern for the USA as a Constitutional Republic, and all other nations wishing to prosper and survive while living in global cyberspace.

Information is the content, the digital substance, of cyberspace. It is information that computers must understand and protect in depth and detail. The Red Queen's power and tragic significance grow as digital convergence accelerates. She, greedily expands physical identities and adopts bad practices on poorly protected binary computers, enslaved to dictatorial operating systems as the Black

Holes in cyberspace. Using AI's incredible, superhuman strengths, the age of the binary computer must end, replaced by the unassailable science of the Church-Turing thesis.

Public security will only exist in cyberspace with democratically enforced digital freedom and if individual machine commands are scientifically limited to perform specific actions as fail-safe machine instructions. Replacing the superuser and centralized operating system with fail-safe function abstractions shared when needed by secret tokens. The level playing field requires every machine instruction to be equally available to every program, programmer, or language compiler. No superuser can exist. The golden tokens are private digital secrets, only shared with discretion, distributing security incrementally to trusted functions. They remain secret unless shared by design with a trusted partner, a user, a programmer, or another data-tight, function-tight, fail-safe instance of object-oriented code.

These computers require golden digital keys to unlock digital access to the location range, type, and mode boundaries of Capability-based addressing using the immutable golden tokens in the individual machine instructions. This combination detects and prevents every interfering mistake from any source. The RISC instruction cannot reference Capability tokens. They are a protected second data type limited to the six Church instructions that define the λ-calculus function of the namespace. The immutable tokens structure the software into individual type-safe function abstractions. Before referencing the content, programs must unlock access to each approved object using the token's name to target each declarative machine instruction.

Computers and memory systems are hidden physical objects, and digital information is only accessible by Capability tokens that act as keys. It controls computational space as a cellular chain of atomic digital things in private, individual threads, just like the computations of the mechanical age that intuitively began so long ago. Their digital-type boundaries and access modes limit access rights. Following the Church-Turing thesis, the information stored in the cloud is structured, guaranteeing digital integrity. Access lists and centralized operating systems have blind spots to these critical networking details. Only the redesigned Capability-based computers understand type-safe security and the symbols of λ-calculus needed to solve AI-enabled cyberspace problems.

Further, the actions of cyberspace are dynamic and unseen, working at the speed of light. The user cannot control binary program interactions within their own computational space. Users cannot be responsible when these things go wrong. The dangerous, multitasked communications execute without human knowledge or permission. Malware irrationally shares the authority of the logged-on user and the ambient power of the overstretched Turing machine. Once again, this is a design fault, not a user error. Finally, all the shared and exposed binary data in the one-dimensional virtual memory is an architectural issue inherited from von Neumann that is not under user control.

The confused mixture of confidential data, multiple programs, and user authority instantaneously delegated to unknown programs, including dangerous ones, is not by conscious user approval but, as Multics decided, expediently, decades ago. Experts, led by Butler Lampson[59], consciously built this unsafe arrangement into their mainframe toys, led by Multics, operated from locked rooms. Downloading code that hides malware into an identity space by browsing or file sharing from a USB drive, email, or messaging is a dangerous threat. Hidden crimes occur in a flash, and the Red Queen remains undetected for days, weeks, and even years, making backup recovery impossible even for expert users.

Dictators, inhibited by fear of things going wrong and lacking detailed control over individual programs, further limit freedom to retain control against Ransomware takeover. Nevertheless, experts still blame users who discover data stolen or damaged. They shout, *"User error,"* and *"Someone failed to follow best practices."* Computer scientists led the public, society, and the nation into this deadly trap. When a program abuses the ambient authority borrowed from the

---

[59] Lampson received his bachelor's degree in physics from Harvard University in 1964 and his PhD in electrical engineering and computer science from the University of California, Berkeley, in 1967. During the 1960s, Lampson was part of Project GENIE at UC Berkeley. In 1965, Lampson and Peter Deutsch developed the Berkeley Timesharing System. He was a founder at Xerox PARC in 1970, where he worked in the computer science laboratory. In the 1980s, Lampson joined Digital Equipment Corporation, and later for Microsoft Research. Lampson is also an adjunct professor at MIT.

identity granted in blind trust by experts, it is willful. The experts, not the users, should pay the price and be held accountable. Investigations would lead to prosecutions and trials if this were the aerospace or the automobile industry.

The irreconcilable blunders made when computer science overstretched the Turing machine, added shared virtual memory, and used privileged operating systems with incomplete and unsafe identity-based security now haunt binary cyberspace. The Red Queen hides behind every innocent step. Passwords and usernames are inadequate; they open access to enormous digital space with amassed confidential information, divergent authority, and grouped but divergent interests. Bad and good programs eat at the same table, sleep, and work on the same machine, share the same digital living space, use the same file storage, cohabit on the same desktop, and share the same corporate network.

Good and bad programs coexist with nothing to prevent interference from digital theft to digital destruction. This identity-based access control model[60] may mirror the Cold War mainframe era, but it does not meet the needs of cyberspace and the mash-ups of an endlessly expanding, AI-enabled society. Networked computers must understand cyberspace, not physical space. Instead, software best practices must be guaranteed by the computer hardware, not based on blind trust and an assumption of following best practices. Automatically, Capability-based addressing enforces the type-safe modularity and the laws of the λ-calculus, creating a scientific Church-Turing machine. This flawless configuration guarantees software reliability by enforcing fail-safe, type-safe, data-tight, and function-tight computer instructions. Using both sides of the Church-Turing thesis, implement computer science scientifically and in full to the endless benefit of humanity, eternally committed to universal cyberspace.

By 1965, when computers began to network, the trust assumptions of physical isolation were already being questioned. The belief that IAM would work beyond time-shared mainframes was naïve, understood by Jack Dennis and Earl van Horn, who proposed Capability-based addressing as a hardware technology at MIT.

---

[60] Identity-based security, Wikipedia.

However, university and industry leaders were unwilling to give up their advantage of dictatorial control. But networking amplified their mistakes that continue to grow, now threatening individuals, society, the nation, and democracy. Indeed, the progress of civilization is at risk. Cybercrime will always be commonplace with binary computers because they are digitally promiscuous, sharing equipment and using unprotected digital assets.

Trusting in best practices instead of engineering a safe, scientific solution is a dereliction of an engineer's duty. It is unacceptable in any occupation. Every other branch of engineering takes public safety seriously. The public good is a moral and legal responsibility. Companies selling unsafe cars that explode in flames, jet planes that fall from the sky, bridges that collapse, or buildings that crumble face trials for ignoring standards and shameful disregard for public safety. The public depends on professional standards and ethics to resolve things they cannot understand or control themselves. Disregarding public trust should require severe punishment. The protection of civilians is a moral issue for every professional engineer, except (it seems) for computer scientists.

Implementing digital security requires digital boundaries, which are vital for the public good. Just as the pillars of a cathedral have engineered properties, likewise, the cornerstones of computer science expressed as the Church-Turing thesis hold firmly onto mathematical science to ensure the public good. It is an ethical requirement to do no harm. At first, computer science was a game, a designer's joyride, but no more. A convenient choice made for first-generation computer science became today's microcomputers, but these computers regularly place the public and the nation at risk.

Discarding the one-dimensional, mutable binary computer where everything is just changeable data is critical. Using the digital ironwork of Capability-based addressing recognizes and reinforces the multidimensional atomic nature of information and implements the λ-calculus. In-depth and detail, the cockpit of a fail-safe, type-safe, data-tight, and function-tight computer executes both sides of the Church-Turing thesis. Information as a structure is distinguished, recognized, and followed by the simple laws of the λ-calculus. Distinct namespace subsystems populated by type-safe function abstraction define Alonzo Church's vision of cyberspace. Computations occur as distributed, parallel threads, sheltered, evaluated, and defended independently, atomically, and automatically by Capability-based addressing and six λ-calculus machine instructions in a Church-Turing machine.

In this way, every named digital object is unique, bounded, and protected dynamically, organically composed in a functional namespace. Cyberspace is secured atomically and dynamically for the mash-up as an active digital organism populated with the individual digital function abstractions of imagined digital creatures needed to serve individuals, businesses, and society's needs. As in biological life, animals evolve gracefully as the environment changes and the DNA responds. Nothing frozen in the past survives. Everything is atomic and cellular, mechanized by a class hierarchy, assembled from cells, and programmed by nature as cellular function abstractions. Each abstraction defines a class of functionality constrained by the laws of the λ-calculus in a need-to-know namespace hierarchy of atomic relationships. The namespace defines the DNA of a survivable software-powered application.

Fine-grained information protection prevents cybercrime, replacing shared binary computers on a grand scale. Attacks and interference are recognized and stopped by atomic digital boundaries. Computation steps are individually checked for digital interference, preventing damage in advance. Secured by Capability-based addressing the λ-calculus vision of cyberspace, it is scientifically protected and programmatically directed by fail-safe program steps. All activity is information-driven. Virtualizing natural items like people, identities, and computers is secondary to atomic and cellular information-based computation.

It is information in cyberspace that is most important. Function abstraction starts with data organized by the λ-calculus, not by time-sharing operating systems or page-based virtual memory. The stereotype of physical computers, memory systems, and individual users as virtual machines was a beginner's mistake that created the Red Queen. It is in the national interest to correct this mistake and purge the Red Queen from cyberspace. Cyberspace belongs to everyone. It is not an industrial dictator's cloud or a computer but the extension of global civilization. Like the abacus, perfection will last forever. As such, universal trust must be scientific and provable. For this, computer science must obey the laws of nature, not commercial preferences or industrial dictators. Criminals and enemies too quickly sabotage digital life, and the nation suffers.

THE SILK ROAD

Digital Equality, Individual Liberty, and Civilized Justice

The Church-Turing thesis builds software as atomic function abstractions. From top to bottom, a private application namespace defines the symbolic, object-oriented machine code encapsulated by Capability-based addressing. Functional correctness stretches uniformly across cyberspace, because namespaces abstract everything as flawless, fail-safe object-oriented software. It is the design of a future-safe *Dream Machine* adapting computer science, as demonstrated by the PP250, to include the latest semiconductor technology with the best functional programming capability. A *Dream Machine* supports anything one can imagine, programmed by self-explanatory machine code. Devoid of undetected malware, minus the centralized superuser, and without the patchworked network fragmented by branded operating systems, every dream is realized symbolically and reliably as atomically networked virtual reality.

If needed for a smooth migration, the software abstractions include every backward compatible function invented for binary computers to port an existing application from a frozen binary computer to an ever-improving *Dream Machine*. It includes abstractions for page-based virtual memory and the range of centralized operating systems, efficiently reimplemented using object-oriented machine code.

These abstractions smooth the transition from today's binary computers to the future safe, fail-safe version of cyberspace. The software abstractions of the *Dream Machine* are a better technology for backward compatibility because the hardware is no longer frozen. The approach not only allows existing applications to run on the new hardware, but the new hardware immediately resolves malware and prevents ransomware since there is no superuser to usurp. Adroitly removing the Red Queen serves cyber-dependent nations' short- and long-term needs.

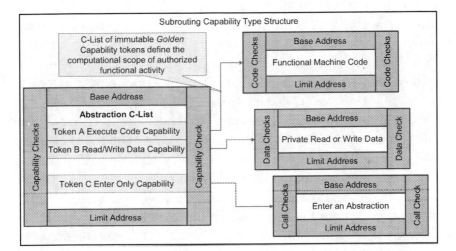

Figure 6. A Capability Limited Function Abstraction
(from the PP250 introduction)

Removing the superuser and virtual memory eliminates binary computers' opaque, monolithic complexity. Digital boundaries support new applications using the full power of networked λ-calculus and the digital voids where malware festers disappear. Cyberspace works far better without any physical constraints interfering with the abstraction process. The physical structures hurt the network abstraction, while the dictatorial operating system blocks networked λ-calculus. Critically, Capability limited function abstractions cannot interfere with one another either accidentally or deliberately.

Delegating the logical to physical translation to object-oriented machine code, the symbolic names of a namespace are, at the same time, the golden tokens of Capability-based addressing and the names of object-oriented software. Capability tokens define the trade routes of cyberspace and make the object-oriented machine code intelligent and readable. At the same time, Capability-based addressing is elevated from a practical memory management scheme to a universal engineering mechanism for many things at once. Capability tokens are readable names and immutable digital keys that can, for example, be secret passwords easily checked by entry guards to top-secret objects.

The PP250 recognized four types of Capability tokens. As *keys*, they unlock access rights to arbitrary digital objects as type-safe memory segments. As *outform* tokens, they represent networked objects. When defined as *passive* tokens, they work as immutable digital values, perhaps an entry password. As a *void* token, the namespace store manager can instantly release resources like storage without first garbage-collecting existing keys.

The PP250 was the first Capability-based computer to achieve fail-safe software reliability needed today for AI-enabled cyberspace as a safe public utility. The computer required just two basic machine types, one for Capability tokens and another for binary data. Each type used different machine access rights: *Read, Write,* and *Execute* for binary data tokens and *Load, Save,* and *Enter* for Capability tokens, with distinctly different microcode machine instructions for each type.

The instructions use the Capability mechanisms as computational guard rails, keeping program executions on track following a hierarchy of functional nodal structures. These are the individual digital details of the namespace, including outform tokens to connect to approved networked objects. For completeness, the standard Turing or RISC instructions also use tokens defining terms referenced by names and offsets instead of a shared physical location.

Binary data and Capability tokens are different machine types created for digital security reasons. The mutable binary data is only for the RISC instructions, while the immutable Capability tokens are only for the security machine, the second half of the computer, a Church machine. Binary data and Capability tokens are like Rudyard Kipling's East and West, *"...never the twain shall meet."* As a result, the RISC instructions cannot access capability data, guaranteeing they remain immutable.

In Figure 5 above, three examples of Capability tokens exist. Capability A supplies *Execute* access right to a block of object-oriented machine code. Capability B supplies *Read* or *Write* access to a block of binary data. Capability C supplies *Enter* access to another digital black box, a function abstraction of any complexity or location in cyberspace. Each black box is an atomic function abstraction with the same general data structure. In a computation, any object-oriented machine code can call any, in scope, Capability token with *Enter* access to safely activate an engineered subroutine call approved by the namespace hierarchy.

Individual threads weave their computational path as silk roads through the namespace. They use a hardware-accelerated, last-in-first-out (LIFO) thread stack as a breadcrumb trail of subroutine calls that unwind the abstractions held in the thread as results mature. The computations are instances of the machine-enhanced Thread class. The functional namespace's immutable keys and the typed information blocks link the function abstractions. Threads time-share the hardware using one of the six Church instructions implementing the private threaded computational model of the λ-calculus namespace, refined by the abacus. The six capability instructions are explained in the PP250 documentation summarized below and in full detail online.[61]

Λ-Calculus Church Instructions:

- **Swap**: New Independent Namespace is Loaded
- **Change**: Suspend Active Thread & Activate Another
- **Call**: Call a Function Abstraction & Push Status on LIFO stack
- **Return**: Pop Abstraction from LIFO stack
- **Load**: Token Access Rights Cashed in a Capability Register
- **Save**: Capability Token Saved to an Accessible C-List

Figure 7. The six Church instructions for λ-calculus computations (PP250 examples)

How an imperfect world of harmful software, human sins, and deliberately dangerous AI coexist with the rest of cyberspace depends on how well Capability-based addressing frames a

---

[61] PP250 Instruction Set
https://drive.google.com/file/d/1EUcG7mTtdko1Kc9Q_PwGG0LiHwLBxHwP/view?usp=drive_link

λ-calculus namespace in endless cyberspace. Names in a namespace symbolically define applications and function abstractions in a more extensive network of many independent application namespaces. The digital modularity of the namespace depends on Capability-based addressing to preserve and protect the digital type and access rights to each object. The tokens are immutable access keys to a unique digital object in the namespace.

The execution threads and various atomic functions use the six Church instructions to translate the tokens into accessible digital objects using the protected namespace table defining the application. The strict separation between the Church and Turing instructions guarantees that binary data and Capability tokens are independent. Thus, only one secret function abstraction, hidden deep in the namespace, prints new Capability tokens for the memory manager abstraction and must be trusted. It involves a few straightforward lines of fail-safe machine code. Far more dependable than the globally shared superuser and impervious to known and unknown malware.

By preserving the information details, the namespace prevents corruption by malware and asserts privacy through the secure ownership of individual capability keys. Object-oriented, type-safe digital boundaries enforce the need-to-know namespace hierarchy as the rules of computation defined by the atomic structure of the information. Unique names in nodal lists, called C-Lists, define a hierarchy of locked digital boxes. Machine code names target each instruction using the immutable digital ironwork of Capability-based addressing. Now, programs gain access to needed atoms and cells in threaded cyberspace. Everything is safely locked up, hidden from hackers, spies, enemies, and thieves, and gifted the added power of functional, readable, comprehensive machine code.

The structure of a functional namespace has nothing in common with the natural world boundaries of a physical computer or even a virtual machine. The study of cyberspace is the science of concurrently processing digital information distributed anywhere throughout independent networks of application-oriented digital abstractions. Processing and protecting cyberspace's wisdom requires the detailed integrity of every atomic element of every individual function abstraction. A physical machine, even one abstracted as a virtual machine, is irrelevant and dangerous in the revolutionary context of malware and AI-enabled cyberspace.

To journey safely through this universal digital medium of programmed software requires innovation to protect every cyber-dependent citizen and every democratic society simultaneously. Monolithic Virtual machines built by experts in one of many branded alternatives only work for experts. They have no modular lines of defense to protect the digital knowledge of cyberspace. Their opaque natures have little in common with other branded applications, making maintenance even more complex. Instead, the engineered digital details are improved when security aligns with in-depth software functionality. Using Capability-based addressing for transparent component reliability highlights the focus on software modularity and individual component reliability (MTBF), exactly like every other engineered feat of history.

Even so, to kill the Red Queen, the medium of cyberspace must follow mathematics as a uniform, universal medium of science. The information galaxy is an ever-expanding universe of data that cannot be held or partitioned by static physical characteristics as seen in the natural world. It is the function abstraction of λ-calculus as a mathematical science, detailed and uniformly equal in importance. Cyberspace cannot include monolithic operating systems, and shared virtual memory cannot support the mash-up of deliberately programmed, unidentified threats that tilt the scales of progress against citizens and a free society.

Cyberspace is a dynamic knowledge space defined by interacting function abstractions of active λ-calculus namespaces. The *Dream Machine* enforces the λ-calculus by dynamically binding type-safe machine instructions to the named procedures of the matching type. Thus, the golden tokens and the digital ironwork of Capability-based addressing prevent bugs and malware while blocking terrorists from entering the cockpit of the computer and enabling functional expressions to reach across the world safely.

When encapsulated as precise functions by Capability-based addressing, the bound commands cannot stray from the programmer's path, limited by the guard rails of a need-to-know namespace hierarchy. Now, each binary program works like the uncomplicated Turing machine as the λ engine in the λ-calculus, as Alonzo Church and Alan Turing first imagined in 1936. Each atomic digital object is a named function as a designated type of abstraction in a named λ-calculus namespace, computed independently in cyberspace.

Like the abacus, the slide rule, and Babbage's *thinking machines*, digital computers using capability-protected software exhibit the atomic behavior of the λ-calculus as a clockwork machine of hardware and software. These scientifically engineered designs have actively measured reliability, pinpointed by Capability-based error detection.[62] The objects with the worst mean time between failures (MTBF) are the top priority for engineered improvement. These are not necessarily hardware components. They are software objects defined by the golden tokens that pinpoint errors of any kind. The mechanical and digital rules of scientific conduct keep dynamic computations on functional rails for high-speed calculations of any digital size. These repeatable, reliable actions process atomic information as a type-safe, function-tight, data-tight, fail-safe cyber-ship. A fail-safe *Dream Machine* built to pioneer unexplored λ-calculus computations across cyberspace. The *Dream Machine* is an updated cyber-ship that safely takes civilians anywhere in cyberspace. Without fail, it returns them with their digital assets unharmed. These computers enforce type-safe, atomic, and fail-safe cellular computational threads of data-tight, function-tight information processing.

However, cyberspace is far more significant to society and the future of life. For example, as previously discussed, dictatorial computers limit individual freedom and the freedom of speech demanded by the First Amendment.[63] While binary computers upend democracy, the *Dream Machine* is scientifically neutral. Preserving the Constitution is vital if the USA and other nations are to coexist meaningfully as a safely functioning global cyber society. If not, industrial dictatorship, rampant cybercrime, and

---

[62] The Clockworks of Computer Security
[63] The First Amendment (Amendment I) to the United States Constitution prevents the government from making laws that regulate an establishment of religion, or that prohibit the free exercise of religion, or abridge the freedom of speech, the freedom of the press, the freedom of assembly, or the right to petition the government for redress of grievances. It was adopted on December 15, 1791, as one of the ten amendments that constitute the Bill of Rights.

international conflicts will run the world into the ground following selfish preferences.

The Constitution of cyberspace is as critical to the progress of humanity and civilization as the Constitution is to the United States of America and America is to the world. As computer-dependent society grows ever more dictatorial, the same human problems resurface that existed in the British colonies in 1776. It is dire because society also suffers from undetected and unpreventable foreign interference by spies, criminals, and enemies reaching internationally and unobserved through universal cyberspace. Every organization uses dictatorial binary computers to spy on and corrupt the affairs of society to meet selfish goals. The *Dream Machine* includes watermarking the golden tokens to trace changes and detect forgery.

Society is vulnerable because corruption originates globally, forcing suppliers to increase their power to restrain crime and hold onto their market share. This self-serving cycle places the industry in charge of law and order because binary computers only use the bottom-up half of the Church-Turing thesis. Yet, computers are the foundation of cyber society. This path leads downhill to unstable dictatorships and increased crime, with one mode for enslaved, powerless digital users and another for authoritarian administrators.

The superuser always inherits privileges denied to ordinary users, limiting digital life's free and happy fulfillment with no means to resist. When deceived or disabled by malware, the operating system is usurped and replaced by an unapproved alternative.[64] Computers must be fail-safe to prevent this catastrophe. Corruption by computer science cannot bypass law and order or squander the human sacrifice the founders made.

When the relentless force of virtualization focuses on balancing the alternative sides of the Church-Turing thesis, virtualization focuses on information instead of equipment, and structured digital security occurs. The λ-calculus clarifies the typed boundaries of every named digital object with properties. These properties guarantee type-safe functionality, digital integrity, individual privacy, software reliability,

---

[64] Joanna Rutkowska is known for researching stealth attacks including the Blue Pill and Evil Maid attacks. She proposed Qubes OS as a countermeasure. Rutkowska continued to focus on bottom-up security.

namespace security, and the reliability of each digital function. Plus, the efficiency of the λ-calculus speeds results, removes baggage, and reduces overheads. The essential mechanisms in this atomic, scientific world are the clockwork rules of the λ-calculus interactions, enforced by Capability-based addressing through object-oriented programming. By default, the digital ironwork of Capability-based addressing implements software best practices.

The sound engineering of the universal clockworks of scientific λ-calculus achieves qualified, engineered software reliability. As Babbage demonstrated, resolving human failures requires science to level the field of play. It needs the perfect mathematics of his *thinking machine* or the golden tokens and digital ironwork of the λ-calculus enforced by Capability-based addressing.

Binary computers ignore the structure of digital data as atomic and cellular information structures. Turing's reel of paper tape was just a linear program of codes. First, shared memory and then virtual memory inherited Turing's linear physical addressing mechanism, but Church's λ-calculus is programmed symbolically, confining the linear binary programs to well-defined logical objects. Accessing shared virtual memory as pages in a large book on loan to every program in the compilation uses unchecked binary pointers. Every program shares the same virtual memory as a linear address space, easily damaged by simple pointer errors or deliberately sabotaged by malware downloaded into the virtual machine. The compiled data executed by a binary computer lacks adequate cross-checks to prevent undetected damage.

Virtual memory is an unsafely engineered, misguided shortcut that solves the wrong problem. Page after page of unguarded binary data all mashed together across virtual memory. The binary information is indistinguishable binary data codes, programs, text string, and numbers only understood by the offline compiler, some related code created by the compiler, or, at times, the operating system. It is decoded and understood as a digital maze by delicate procedures that must match every detail and mindlessly trust best practices apply at every step in every program, even to downloaded code. Worthy and unworthy software shares this dangerous digital void and related assets without severe constraints or fail-safe protection.

Virtual memory is easily damaged, like popular books in a public library. Shared ambient authority in binary computers is public and cumulative instead of private and individual when using the λ-calculus. Sooner or later, programs silently interfere with one another through shared memory. In the digital void, law and order are missing. By crafting the values of unchecked binary pointers used by machine code to navigate the shared memory, malware can spy, steal, and destroy the hard work of nations, businesses, and individuals.

As explained, computer science began with the abacus, the wooden machine of rails and beads invented at the birth of civilization, and prosperity grew because everyone excelled at arithmetic. Amplified by the powers of mathematics using Oughtred's slide rule[65], engineers excelled, evolving the Industrial Revolution, leading Charles Babbage to invent his difference engine, dubbed by Ada Lovelace the *thinking machine*.[66] Unlike dysfunctional binary computers, the easy-to-use mathematics of a *Dream Machine* serves society equally and faithfully for everyone. Individuals are empowered and intellectually extended using the personal, private threads that digitally power the human condition instead of criminals and dictators.

After the λ-calculus codified the science as the Church-Turing thesis in 1936, the PP250[67] demonstrated fail-safe, function-tight, data-tight, type-safe capability-based computer science in 1972. It pioneered the flawless bottom-up, top-down architecture for computer science. In every case, these dynamic, diligent, reliable computers stretching back thousands of years physically defend the functional boundaries of every atomic mathematical type. Like the abacus, PP250 implemented private computational threads as atomic cellular executions.

These atomic algorithms link dynamically as in mathematics, using a frame for the abacus, grooves for the slide rule, and hard digital tokens for PP250. For Alan Turing, one procedure was stored individually on one paper tape. It formed a functional, stand-alone computation used successfully by the abacus and the slide rule as private individual threaded calculations.

---

[65] William Oughtred, Wikipedia.
[66] Difference engine, Wikipedia.
[67] Plessey System 250, Wikipedia.

Unfortunately, von Neumann, in a hurry to be first, overstretched Turing's idea, creating centralized shared computations that overlap and lack engineered law and order. He should have known better. He knew Alonzo as a colleague and met with Alan Turing several times when they discussed implementation options before Turing finished his doctoral thesis. Turing then returned to England to work on code-breaking for Ultra[68] during the Second World War.

Ever since, designers have enjoyed inventing and reinventing binary computers. They play God in creative ways to redefine computer science using the centralized dictatorial operating system. They dismiss concerns about malware overturning the operating system. But time proved none of the many alternatives is stable. They cannot meet the goals Buttler Lampson claimed at the SOSP78 summit meeting at Perdu University. Butler disparaged and detracted the qualified results of PP250[69] field tests and used his stature to dismiss Capability-based computers. The operating system was far more interesting to attempt. However, hacking and cybercrime became rampant once the internet took off, and cyberweapons continue to grow stronger with new AI-enabled software and frozen hardware from the past. Foreign enemies are now writing breakout code for their AI-enabled malware attacks.

Charles Babbage and Alonzo Church would be distressed by the uncertainty and unpredictability of undetected malware. It is all caused by digital insecurity in shared binary computers. Conversely, government agents, warriors, hackers, criminals, terrorists, enemies, and spies could not be more delighted. Adding AI will only worsen events until mainstream computer science adopts the complete bottom-up and top-down architecture of computer science.

All the binary baggage is simply misguided overhead dangerously balanced on digital tricks and opaque hardware settings. It is a calculated choice made by industry dictators to retain power and

---

[68] Ultra was the designation adopted by British military intelligence in June 1941 for wartime signals intelligence obtained by breaking high-level encrypted enemy radio and teleprinter communications at the Government Code and Cypher School (GC&CS) at Bletchley Park.
[69] Panel session led by Bob Fabry included Lampson and Hamer-Hodges to review Capability Based Addressing. Lampson's claims were never published, but my report can be found here.

grow captive markets. Nothing invented, from software patching to extra security control lists, read-only memory units, and software monitors, solves the problem of binary exposure. There are no software best practices to correct von Neumann's mistake. Binary computers are incompetent, simply incomplete computer science.

Using the *Dream Machine*, the science of the λ-calculus stretches uniformly across computer science to the hard edge of cyberspace. Atomic modularity isolates every access to the computer's memory. The digital voids, the privileged hardware, and the proprietary operating systems disappear. The laws of the λ-calculus apply universally throughout cyberspace, removing every trace of malware and the unfair powers of centralization.

Proprietary baggage discredits computer science. It includes leaky firewalls, imperfect antivirus scanners, ineffective access control lists, frustrating numbers of passwords, worthless certificates, constant upgrades, and patched-up software only delivered after successful attacks. It is all unsound and unsatisfactory. Unscientific branded baggage is delivered too late and too little to save the future. These practices cannot resolve the science missing from binary computers. It is worth repeating that von Neumann's overstretched architecture is at fault, and there is nothing that software can do to patch over or hide this hardware defect of centralized sharing.

Black-Holes in cyberspace grow bigger and bigger, and the overheads are more suffocating and debilitating. At the same time, AI-enabled attacks by friends and foes will increase in number, strength, and stealth to penetrate the digital core in unacceptable ways and numbers. The only solution that works is the integrated ideas of Alonzo Church and Alan Turing. The disciplined encapsulation of algorithms as object-oriented software protected as a multiprogrammed computer network by Capability-base addressing enforces the laws and order of the λ-calculus to the hard edge of cyberspace.

The world's dependence on AI-enabled software and cyberspace as a Public Utility makes the infidelity of binary computers unacceptable. First-generation binary computers have changed from toys into demons. Only the *Dream Machine* guards the individual users of cyberspace. Nothing less will do. Already, the peculiar dangers of malware escape into the natural world as the Red Queen attacks and destroy anything targeted, even machines on the far side of the world. Automated AI bots already outsmart human behavior using social networks and deep fakes, rewriting history and discombobulating truth with lies. Soon, AI will create the unperson.

Artificial intelligence is taking over. AI cannot run binary computers as private, industrial dictatorships and usurp the nation's laws and order. It will irrevocably change human rights as cyber war between industrial dictators, criminals, and enemies provokes the Red Queen to run the world. Nurturing young life is vital because it is fragile; if not, the weeds of malware and digital corruption will strangle the Constitution in cybersociety.

It is all the unsettling consequence of unsafe, unscientific digital convergence. The Government must act to secure the institutions of democracy. Undemocratic digital powers usurp the common good through crime and dictatorships that undermine and overtake democracy, human values, and national identity.

As the new digital order attacks the digital future, think about how to protect and preserve the founding assets of the nation, written in the Constitution. What do freedom and privacy mean in cyberspace when AI runs everything? The Constitution and the Bill of Rights must cover cyberspace, or AI-enabled cyberspace will replace the Constitution. Does protecting an individual's right to digital self-defense extend into cyberspace?

Only capability keys manifest a solution that equates to carrying arms in cyberspace. As lawlessness spills out of cyberspace, it polarizes every computer-connected home. When nothing is safe, the long-term consequence is a world run by Big Brother, a dictator far worse than King George III. Debating amendments to the Constitution[70] to cover cyberspace will extend the laws of the land to avert subversion by the corrupt forces of the Red Queen and hasten the *Dream Machine*. The Blueprint for an AI Bill of Rights is an essential step in the right direction[71].

---

[70] Constitution of the United States, Wikipedia.
[71] The Blueprint for an AI Bill of Rights

THE AI COCKPIT

Fail-Safe Computations

The cockpit of computer science is where all the digital actions take place. Everything good or evil starts here. Programs only execute machine code directly or indirectly defined by programmers; even AI is achieved step by step in the cockpit of a digital computer. The pilot program selects every step in a binary instruction based on prior conditions. These instructions expend electrical energy and change the state of the digital machine. The result transforms into digital actions as the algorithm runs. But the tilted playing field of dangerous binary computers enables malware to start here, on the edge of cyberspace, as software transforms to create virtual reality.

The inequality between users and superusers limits what machine commands can and cannot run. Beyond this crude check, intended to protect the branded hardware invented for virtual memory from attack, the software is mindlessly trusted. AI, malware, and programs use the binary instructions in any permitted way, including misusing shared memory and any other shared equipment. The superuser exposes everything that controls the computer to a deliberate or accidental attack. The malware can spy, steal, demand a ransom, or cause a catastrophe by targeting the misuse of shared memory or attached hardware.

Mindlessly trusting users and superusers is why the playing field is tilted. Malware takes advantage by doing unexpected harm. One wrong instruction can change the pilot program and lead to the total subversion of the operating system. These attacks are like the 9/11 terrorists breaking into the cockpit of unsuspecting passenger jets. Under AI control, they can do the equivalent destructive harm to computer-dependent society as terrorists did when they brought down the World Trade Center. These attacks are fascinating and researched by experts, including Blue Pill and Evil Maid, exposed by Joanna Rutkowska.[72]

Joanna's solution, Qubes OS, is a countermeasure that physically isolates an adjunct computer's operating system and memory system. However, the problem is logical; physical solutions cannot detect logical errors in a binary computer. Computer science depends on protecting the functional details of well-written software as fail-safe commands. By following the science of the λ-calculus, the machine instructions are validated and made fail-safe. Software, protected symbol by symbol, is the solution that detects every logical error and, when fail-safe, prevents all forms of undetected digital cybercrime. It is where Capability-based addressing steps in. This enhancement adds symbolic tokens to the machine code with two significant advantages.

First, tokens add the power of functional programming to the machine code; tokens represent digital functions as black-box objects. A black box can be as small as a read-only or read-write binary data block. It can also be an executable block of code, a function abstraction, a device driver, or a digital type like a number or a character. As an abstraction, a black box can also be an immutable program constant or a complex type. Furthermore, a black box can be large, even huge, as a subsystem or another namespace, locally or remotely found in universal cyberspace. The mechanics remain the same in every case, and the tokens pass as variables between abstractions.

Second, thread objects abstract private computations as individual, protected calculations, as demonstrated using a slide rule, an abacus, or Babbage's *thinking machine*. A personal analysis may use a shared service but requires no centralized operating system. Converting cyberspace from dictators with absolute power to a threaded democracy applies mathematical and logical equality of science through symbolic abstractions. Ada Lovelace identified this

[72] Joanna Rutkowska, known for stealth attacks including Blue Pill and Evil Maid attack, proposed Qubes OS as a countermeasure. Rutkowska continued to focus on low-level security.

profound vision of the distant future in the 1840s. Three simplifying hardware steps turn opaque centralization into transparent distribution, in addition to modular security, data privacy, networked security, and fail-safe concurrency.

First, remove all the proprietary baggage invented by suppliers over the past half century and return to the simplicity of the first Turing machine that only processed one algorithm at a time. Second, add six λ-calculus machine commands called Church instructions.[73] These machine instructions programmatically guarantee function-tight, data-tight, and type-safe access to cyberspace. They control the navigation of the λ-calculus threads through the namespace hierarchy. Only local tokens are accessible once unlocked by the Load Capability instruction. The individual capabilities manage digital security incrementally as an approved access right for use in the cockpit. The redesign includes retargeting RISC instruction from linear addressing to specific λ-calculus-named objects in an application namespace. Finally, add Capability addressing in hardened micro-code for the Church instructions as a copilot to cross-check the active instruction against a specific, already unlocked λ-calculus token. The gain in hardware simplicity and software transparency is priceless just by removing the outdated baggage of the binary computer.

As explained previously, every object can be protected when scientific symbols express connectivity between objects. Moreover, symbols that represent functions can pass as variables. This software, like Ada Lovelace's Bernoulli abstraction, survives indefinitely. Thus, over time, software development costs trend to zero, along with undetected cybercrime. But most importantly, the existence of centralized dictatorships in cyberspace evaporates. Each functional namespace is self-standing. At the same time, data privacy is automatic, and the symbolic power is incremental. The λ-calculus can reference objects on the far edge of cyberspace, unblocked by removing the operating systems. All power is distributed to the namespace tokens, now in the hands of programmers, users, and ultimately *"We the People."*

---

[73] The Church instruction for PP250 called capability instruction are explained in the PP250 Instruction paper.

Digital democracy begins with the flawless implementation and private modularity of the symbolic λ-calculus instead of shared linear addressing and centralized compilations. Capability-based addressing checks in depth and detail every atom of digital modularity by type, scope, and access modes. Respecting each scientific symbol is vital, guaranteeing results remain error-free. Finally, and most importantly, the structured namespace hierarchy keeps unknown and unapproved objects, including AI malware, out of the cockpit.

The clockworks of the λ-calculus microcode navigator sit as a copilot beside the conventional RISC machine pilot. Together, they cross-check each other and enforce every dynamic action against the programmer's namespace plan to set and check the object-oriented guard rails for each named target. These digital boundaries enforce each case (active, passive, outform, null), both types (data or capability), every location (local or networked), and two modes (read or write).

The mechanisms guarantee the high-speed guardrails built from the digital iron defined by the immutable tokens of Capability-based addressing applied in full depth and detail. It allows programs to strip down like racing cars, removing redundant in-line checks like the inefficient operating system security monitors. Capability-based addressing performs digital security faster and far more effectively. Finally, the λ-calculus prevents malware from accessing the cockpit and flying results off course. The dangerous digital terrorist, zero-day attacks, and 9/11 cockpit attacks that lead to Ransomware, AI breakout, or other catastrophic disasters stop because unknown software does not exist in the namespace.

Like the aerospace industry, computer science needs a terrorist-proof cockpit with a copilot to act as the navigator. The pilot, as first designed by Alan Turing, executes stripped-down binary instructions, and the co-pilot executes the navigation instruction as defined by the laws of the λ-calculus. The computer is well balanced when the top-down science of the λ-calculus encapsulates the bottom-up digital mechanics. Computer science becomes a *Dream Machine* with a level computational field of play. Only then is computer science safe for public service on the grand scale required by global cyberspace.

The λ-calculus copilot navigates a preflight plan filed as an application architecture by the programmer, aided by the program user. This engineered plan, by definition, excludes all unknown programs, malware, and unknown foreign objects. Because programs now follow the function plan of approved destinations, filed as a hierarchical namespace, the rules of λ-calculus modularity implement digital guard rails. The guard rails encapsulate each private computational step as a fail-safe machine instruction—a private calculation protected from malware and AI breakout in terrorist-proof threads of safely secluded computations.

There is no virtual machine or a centralized operating system, just individual event threads in distributed, asynchronous, but concurrent networked computations. Each privately threaded calculation is unique, independent, personal, and fail-safe. Each step is a fail-safe machine instruction cross-checked on the spot by the encapsulating control of the λ-calculus namespace. The binary pilot plots a path limited by the hierarchical routes of the linked tokens strictly checked by the λ-calculus copilot. It brings the full power of the Church-Turing thesis into play as the perfect level playing field of computer science. The cockpit crew understands λ-calculus machine code (λ-MC), and instead of unchecked and overpowered RISC instructions, the λ-calculus limits activity to the strict mathematics of the namespace. In a Church-Turing machine, the λ-MC instruction includes a set of lightweight RISC functions. Each is data-tight, function-tight, and fail-safe, logically and functionally cross-checked by the laws of the λ-calculus that defines mathematical computations in the simplest complete form.

The pilot controls Turing's binary instruction, while the navigator executes six microprogrammed Church instructions. The Church instructions can load a namespace, switch between threads, call or return from a function abstraction, load a function as a subroutine abstraction, and unlock access to a memory segment defined by a Capability token. The instruction caches each golden ticket, location, object machine type, and access rights in a Capability register. Alternatively, save a Capability token into a previously opened *C-List*. The thread caches the unlocked objects in the dedicated machine Capability registers. The crew cross-checks each other to detect every single error and prevent any single point of catastrophic failure.

Effectively, two machines work in harness following the λ-calculus rules of modular function abstraction in the active application namespace. The immutable private tokens of object-oriented, Capability-based addressing, defined by the λ-calculus, replace shared linear addressing. The namespace hardware dynamically binds the tokens to unique digital objects, performing and caching in real time as implemented for the web. However, URL strings are mutable pointers and, thus, forgeable. The critical difference allows networked cyber crimes, including cross-site and denial of service attacks. Preventing these attacks requires computer science, as defined in 1936.

Thus, the navigator enforces the programmer's design plan, detecting every pilot error or malware infection and keeping programs on track. Running programs between guardrails defined by mathematical functions and removing direct access to shared default powers stops all silly, hard-to-find bugs and any attempt at AI breakout. This check and balance automate software best practices and prevents all cybercrime while elevating the computer's functionality with the scientific powers of the λ-calculus and functional programming. These advantages are universal. They reach the edge of cyberspace with the added and far more powerful mathematics of functional programs, as Ada Lovelace demonstrated when she programmed Babbage's flawless *thinking machine.*

The Dream Machine perfectly avoids cybercrimes while enabling the democratization of cyberspace by preventing the dictatorial whims of the computer industry from crushing individuality. Equality and freedom now extend to cybersociety.

Industrial dictators only limit cybercrimes by increasing domination and enslavement, but switching to the λ-calculus removes the shared, overlapping, and default privileges that corrupt binary computers. When misused by AI, these dangerous, unscientific powers can attack every level of digital society.

When both sides of the Church-Turing thesis meet at the hard edge of computer science, the modular laws of the λ-calculus tame AI and all forms of malware. The shared virtual machine, the superuser, and the centralized operating system dictators vanish, replaced by the simplified computational model of individual threads, avoiding the dangerous mistake of blind trust.

The combined work of Alonzo Church and his student, Alan Turing, scientifically defined computer science in 1936, before the semiconductor age, provided the best technology. However, the binary computer in all present forms still ignores the λ-calculus, and the dangerous RISC instructions remain unconstrained. Today, a Dream Machine on a chip could cross-check the physical actions using the λ-calculus to hide the details, creating, as scientifically intended, perfected programmed modularity. Measured by the fail-safe events, the mean time between failures (MTBF) calibrates and prioritizes any needed improvements. Flawless computers that level the playing field of computer science prevent an Orwellian takeover by deep fakes of Big Brother. Most importantly, for the progress of democracy, Capability tokens place digital power in the hands of users instead of dictators for the ever-safe future of AI society.

The λ-calculus is a scientific masterpiece, the soul of functional software, sitting in the cockpit as a copilot and navigator programmed activity. The soul is the hierarchical assembly of immutable tokens defining the trade routes of software. It is the DNA of links in private *C-Lists* to digital objects and object-oriented function abstractions. The digital ironwork of immutable capabilities links function-tight, type-safe, fail-safe, and data-tight software boundaries. It is the only acceptable digital computer where humanity can survive as a stable, democratic, safe, secure, progressive AI society.

All too quickly, as computers changed the world, industrial dictators deliberately froze their binary computers. Decades ago, long before the internet, computers were anchored to the mainframe past through a selfish backward compatibility strategy. [74] The choice protects their markets from disruptive change and outside competition. They justified this because of the enormous amount of money they wasted on the clunky operating systems and the static procedural software of the mainframe age. So, in the 1970s, the first microcomputers, the Intel 80xx series[75], and all others adopted the shared binary architecture of von Neumann and Multics, adding page-based virtual memory. The progressive innovation using

---

[74] Backward compatibility, Wikipedia.
[75] Intel 8008. The Intel 8008 ("eight-thousand-eight" or "eighty-oh-eight") was an early byte-oriented microprocessor designed by Computer Terminal Corporation (CTC), implemented, and manufactured by Intel, and introduced in April 1972. It is an 8-bit CPU with an external 14-bit address bus that could address 16 KB of memory.

Capability-based addressing computers, started by Professor Maurice Wilkes in 1969 at Plessey Telecommunication and then Cambridge University, with CAP, was once again ignored.

The PP250 adapted symbolic addressing to align Capability-based addressing with the λ-calculus, functional programming, and object-oriented machine code. Six new Church instructions deconstructed shared, centralized binary compilations to private object-oriented implementations programmed using the object-oriented machine code. Independently, Object-oriented programming began at that same time in Oslo using the Simula language[76] and later progressed through work at the Xerox PARK research center with Smalltalk.[77] But the PP250 was in industrial service long before any other form of Object-Oriented programming. It was successful for over a decade before Objective-C and C++ changed the focus of mainstream programming languages.

Object-oriented machine code is a far better technology, dubbed at that time for PP250 as Capability-based addressing. Specifically, the PP250 supported engineer software reliability with a Mean Time Between Failures exceeding any computer. As a fault-tolerant computer system with software and hardware, over half a century of fail-safe operation was possible. Designed by Plessey LTD for high-reliability telecommunication networks,[78] the UK Army adopted PP250 for the Ptarmigan mobile switch, which saw service in the First Gulf War. However, without a silicon foundry, Plessey Telecommunication could not compete with the semiconductor industry and against the combined opposition of Multics, Butler Lampson, and the computer industry, all wedded to the binary computer.

---

[76] Simula is the name of two simulation programming languages, Simula I and Simula 67, developed in the 1960s at the Norwegian Computing Center in Oslo, by Ole-Johan Dahl and Kristen Nygaard. Syntactically, it is an approximate superset of ALGOL 60, and was also influenced by the design of Simscript, a free-form, English-like simulation language conceived by Harry Markowitz and Bernard Hausner at the RAND Corporation in 1962.

[77] Smalltalk is a purely object-oriented programming language (OOP), created in the 1970s for educational use, specifically for constructionist learning, at Xerox PARC by Learning Research Group (LRG) scientists, including Alan Kay, Dan Ingalls, Adele Goldberg, Ted Kaehler, Diana Merry, and Scott Wallace.

[78] Plessey System 250, Wikipedia.

Since then, computer science has preserved the backward-facing, privileged, static software dictatorships inherited from von Neumann and the mainframe age. Clunky computers built for simple procedural programs running in eight-hour shifts between cold restarts. However, everything today is networked and dynamic, as needed for PP250. Programs are downloaded by browsing the internet, sharing files, email, and opening message attachments. Hackers and criminals attack the static, outdated binary architecture from the Cold War days on the mainframe. Urgent patches and frequent upgrades try to fill each digital void and logical gap. For binary computers, these rearguard fights will never end. It is why Plessey developed the PP250 to address a new generation of fail-safe, Capability-based computer science. Computers that excel at enforcing the atomic modularity of functional software in global networks.

As with nuclear science and the biology of life, software survival depends on the laws of nature, not the egotistical ideas of von Neumann's and Butler Lampson's followers. Natural transparency is atomic, functional, and scientifically understood. Unskilled users quickly adapted to decentralized, individual, and private digital computations as promptly as the Babylonians accepted the abacus. Opaque, privileged, centralized operating systems are all unstable, and constant patched upgrades expose more problems than are fixed. They never work as intended for the global market that exists today. Instead, decomposing the operating system into function abstractions[79] works atomically, like everything else in the universal science of nature.

Capability-based addressing and the λ-calculus reverse the branded monolithic assumptions of binary computers. Computer software cannot, under any circumstances, be trusted when strangers program software. Credibility for blind trust evaporates. When the binary computer started, the only programmers were staff members, counted as good friends. Those days have passed, and schools teach programming worldwide. Trust is meaningless when applied to anyone, soon everyone.

A Church-Turing machine firmly enforces the laws of modular, dynamic, λ-calculus computations. Even amateur programmers

---

[79] Capability concept mechanisms and structure in System 250.

can be trusted when hardware guarantees software best practices. The PP250 double-checked every action before allowing any read, write, or hidden functions, including the instruction fetch and the Capability token translations. The checks are performed in real-time in the cockpit of a Church-Turing machine to guarantee type-safe, data-tight, and function-tight computations. The fail-safe hardware works beside the pilot and navigator, inspecting every step, most notably every memory access, including those related to the program and the computer machine cycle. Only hardware can guarantee data-tight and function-tight law and order. The scientific enforcement of digital modularity following λ-calculus detects every interfering error in any line of object-oriented machine code. Errors are detected and prevented by Capability-based addressing, resulting in fault-tolerant, fail-safe software achieving decades of malware-free reliability.

In a binary computer, malware deliberately and silently causes trouble. The slightest degree of digital interference is hacking, as every cybercrime starts. It is all driven by the unchecked binary computer's outdated first-generation, single-seat cockpit. The cockpit is where digital voids are blindly trusted and then shared. Centralized, default powers exist that reach into every digital corner. Spies snoop, criminals steal, and enemies attack the virtual machine using the unchecked default, ambient, and inherited authority to topple the unstable superuser.

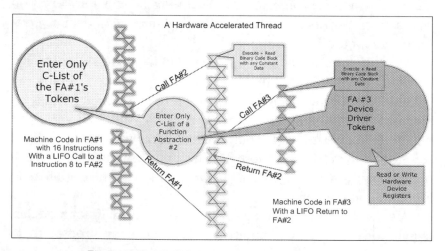

Figure 8. The Call & Return Instructions in a Private Thread of Computation (PP250)

Capability-based addressing enforces the λ-calculus using symbols that tolerate no digital interference. The PP250 gave distinctive functional names to each Capability token to uniquely identify each command's functions and terms. The type-safe digital boundaries tame AI and malware simultaneously by implementing every λ-calculus expression as a fail-safe digital instruction. Programs stayed on track with less code running at higher speeds by working safely within the need-to-know guardrails of the application namespace. Unlocking each term for access is performed by one of six Church instructions. An immutable digital token, standing for the name of the digital object, acts as a term. These tokens confirm the design approval of the given access rights within the digital boundaries. Further, these tokens work with critical computer functions like garbage collection and originator watermarks.

A PP250 treats every programmed action with suspicion, handling each λ-MC instruction carefully to detect and prevent digital infection. Binary computers cannot do this because the imperative instructions are commands performed without any question, allowing malware any desired access to all the shared physical address space. Added programmed checks provided by compilers are unnecessary overheads and too far from the point of action to work reliably. The place for science to apply is in the cockpit. Attacks originate here as instructions execute and the software comes to life, sometimes as dangerous malware and crimes occur. The cockpit is the only spot where computer science is guaranteed or lost for good.

Blind trust always fails because computer hackers exploit the unchecked authority created by senselessly sharing a physical space with strange programs in an overstretched design. Individually, the imperative commands first invented by Alan Turing can do enormous harm. A single binary instruction could wipe the memory clean. The instructions are where hackers and criminals start their work to undermine digital integrity. Thus, machine commands are where digital security and mathematical integrity must come together. The centralized software baggage and added compiler overheads cannot help; only another purpose-built software patch can keep things working, and patching can take months, further degrading performance. Worse, the upgrade may create another flaw or not be installed simply because everything takes time and adds new risks.

Meanwhile, cyberspace rolls downhill, and the criminals, enemies, and dictators steadily win.

Trusting programs to follow best practices is believing in the tooth fairy. It does not work in cyberspace, for the same reasons piling stones in Babylon failed. Compiling software into piles of indistinguishable binary commands and compacted binary data called virtual machines fails today for all the same reasons stone piles died in Babylon.

Imperative binary commands of a RISC[80] computer have unfair, dictatorial powers over the computer. Trusted computers require every machine command to be functionally fail-safe and scientifically constrained. The guard rails of Alonzo Church must be included and constantly checked. In the unused half of the Church-Turing thesis, enforced by capability addressing, every instruction must stay within the boundaries, types, and access modes approved as namespace constraints by the programmers. Furthermore, the PP250 named the immutable tokens logically as readable names to quickly comprehend and navigate the machine code.

Binary computers cannot guarantee life when civilization depends entirely on software automation. Cyberspace is too complex, and the future is too open-ended to rely on the unchecked commands of any binary computer. Instead, confining programs according to the functionality of the λ-calculus guarantees best practices cannot fail by accident or design. The λ-calculus embeds the type-safe goals of each scientifically defined symbol while assembling each namespace, and the digital-iron rails of Capability-based addressing keep the software on the right track. The guard rails act like a train running on iron at higher speeds, pulling heavier loads without unnecessary overheads and always staying within the authorized scope of each function.

The Dream Machine understands the local context of one function abstraction at a time. One atomic node in the namespace hierarchy atomically holds a few local names in a private *C-List* for one function abstraction. Each node hides the access tokens, preventing other nodes from following the need-to-know hierarchy without using the functionality that protects the application namespace as the designer

---

[80] What is a reduced instruction set computer (RISC)? Definition from Techopedia.

intended. It is how the infinite mathematics of Alonzo Church's scientific vision of cyberspace works. Computational threads navigate the namespace by following the functional names as an atomic link, evoking a dynamic interaction between named function abstractions. Symbolic names are how Capability-based addressing protects a type-safe computer memory of object-oriented machine code. Finally, and most importantly for the future civilized AI-enabled society, individuals control the tokens instead of dictators, governments, spies, and criminals. Ultimately, using immutable digital tokens is how modular software prevents AI breakout.

Many more citizens make daily journeys through cyberspace than fly. Society suffers more from computer failures than aircraft problems. Travel in cyberspace demands the same level of federal support as a Public Service as the FAA. It is time to declare cyberspace a Public Utility and create an agency dedicated to the public interest of living in cyberspace for eternity. A Federal Computer Administration should set and check the industry's standards as a critical industrial technology, or the issue of cyber security will only grow. The industry is too self-interested to allow open hardware research and free competition to do it themselves. The frozen hardware must quickly catch up with progressive malware and AI breakout threats. Cyberspace, as the extension of individuals, cannot be compromised or tainted by centralized operating systems for decentralized computer science to succeed.

The Constitution must guarantee individual freedom through a Bill of Rights in Cyberspace. Government action is required to secure solid progress toward cyber democracy and end backward compatibility and digital dictatorship. Governments must clarify the rights of individuals in cyber society as an extension of the individual, not of dictators and their henchmen.

THE BABBAGE CONUNDRUM

The Flawless Computer

After Henry Briggs published his logarithmic data, evaluated by hand and eye, his compilation gifted the static powers invented by Napier to every interested individual. The increased precision over the embedded scales of a slide rule belied any hidden human errors. Nevertheless, the results are unmatched by rods and scales using the human eye. The hidden mistakes in compiled data are always the same, like a pile of dumb stones. Unthinkingly trusted, the embedded errors and interference from wear and tear or outsiders remain undetected, leading to unresolved cybercrimes in today's cyberspace.

These hidden errors exasperated Charles Babbage when he said in frustration to his friend John Herschel, *"I wish to God these calculations had been executed by steam."* The grindingly tedious labor of manually checking tables was one thing. Worse was their hidden unreliability. It set Babbage on his quest to unravel his conundrum by mechanically automating the flawless production of mathematical tables.

When such tables are without error, long multiplication is a sequence of lookup-addition-lookup steps spread across the physical pages of the logbook. Each lookup step binds a data value, expressed symbolically as the exponent of *log x*, to actions in the procedure to compute:

$$log\ xy = log\ x + log\ y$$

The lookup procedure and the pages of tabular data replace the original method of long multiplication. When performed by a binary computer, this process, called data binding, uses unprotected binary values as the data pointers. In this case, the pointers reference cells in a data structure, a digital representation of a logbook table. The software must select the correct data cells to reach a flawless result. But binary data pointers are unreliable, easily corrupted, or accidentally miscalculated, and the data table is opaque.

The binding process ordained by the values of the pointers demands perfect program practices. However, checking every mutable binary pointer adds overhead because the layout of the opaque data created by the compiler is intimately related to the lookup procedure. Changing the structure without recompilation and regression testing the build is impossible, and any undetected corruption eventually causes hidden errors. The changed data map unavoidably changes the code needed to calculate the pointer. Furthermore, compiled data, tarnished by prior corruption or mischievously manipulated by a wrong step, is incorrectly targeted, and undetected mistakes occur. Such errors are far worse than detected errors. Detected errors are always correctable, as learned from teachers at school, but hidden errors escalate into tragedies.

The undetected error is the Babbage Conundrum. The impact of such mistakes worsens, and eventually, for a binary computer, something dramatic like running out of dynamic memory happens, and work is lost. Anyone familiar with word processors knows this event because hours of challenging work suddenly disappear. Over time, patched upgrades have limited these events for an established product like a word processor, but not without substantial unplanned overheads, plus human and financial costs. These setbacks repeat for every new software application release.

Worse, any imported malware within the same virtual machine can deliberately provoke new errors. Hackers do it for fun, but others craft the attack to usurp the system, topple the operating system dictatorship, take control, encrypt the backing store, and demand an immediate Bitcoin ransom. This deliberate cause of havoc has no bounds. Corruption spreads as fraud and forgery substitutes for valid data to trick the users and the operating system. Then, silently replacing the industry dictatorship with a worse, enemy alternative.

But in the twentieth century, the slide rule, more than the logbook, was bread and butter for school teenagers, students, engineers, mathematicians, and scientists. For a subject demanding accuracy, the convenience of the slide rule has practical considerations of physical length limits. Alternatively, the numbers in a printed book serve accuracy well but suffer from systemic, undetected, potentially catastrophic hidden errors. The source of the Babbage Conundrum is a human error caused by an accidental miscalculation, a mistaken typesetting step, or a human print error. Pointer errors and deliberate malware interference cause computational failures, but hidden attacks are endless, including digital wear and tear like memory runout. Every error degrades the reliability of the results.

Removing all sources of digital error is the scientific solution to the Babbage Conundrum. His goal was flawless computer science. A computer with unguarded shared data, such as a logbook or shared virtual memory, runs the added risk of accidental breakdowns from wear and tear and deliberate outside interference. Worse, the assumption of blind trust built into the binary computer exposes the dependent users to an undetected, disastrous catastrophe before the mistake is detected. These failures occur without warning because there is no computational safety net or meaningful guard rails. A second factor must cross-check and double-check every intermediate step toward a correct result. But results in a binary computer only depend upon trusting in unstable, incomprehensible, and swelling lists of conflicting industry and application best practices. Sadly, industrial dictators now depend on users discovering errors.

Criminals search for zero-day attacks that come without even a day of warning. When malware exploits a zero-day attack in a binary computer, it can grab superuser control away from the operating system. It leads to a far more potent attack by importing code from a hidden home base beyond legal reach. These attack options are limitless, driven by skilled gangs from anywhere worldwide. A straightforward slip escalates to total takedowns of the finely but precariously balanced binary compilation where procedural steps must always match the memory map. The virtual machine, including the dictatorial operating system, is a house of cards; one dynamic slip in either a procedure or the data can sabotage the device and remain hidden for months before bringing everything down. It is not a suitable platform for civilized life to live safely as an endless democratic nation.

Babbage proposed, pioneered, and proved the civilized, scientific solution to his puzzle. It is flawless computer science. The abacus and the slide rule, as robust examples of computer science, limit freedom of movement to match mathematical functions. The minimum requirement for flawless functionality is the encapsulated enforcement of type-safe, function-tight, data-tight, and fail-safe boundaries. It is why computers like PP250 added the digital ironwork of Capability-based addressing to define and defend digital integrity and prevent invalid memory access. Capability-based addressing offers an immutable digital framework to protect individual digital objects and to avoid malware entering the cockpit and hijacking the computer. It includes detecting out-of-range pointer errors, fraudulent programs, unhandled program errors, and thread or stack errors crafted by skilled attackers. Indeed, PP250 showed that every code error, including coding mistakes, quickly leads to a functional boundary, type, or access mode error. It closely identifies the problem as one specific instruction in a particular thread. Debugging could not be easier, even for the most significant applications that are no longer opaque and monolithic.

As a working function abstraction, a slide rule shows how separately bound algorithms can share a common scientific framework safe from undetected interference. For practical reasons, length limits accuracy to the most significant digits. However, the slide rule delivered magnificent engineering results for hundreds of years. Samuel Pepys needed ships of the Royal Navy, and the tables estimated the required cubic volume of wood. Stevenson engineered the details of his steam trains. Isambard Kingdom Brunel built beautiful suspension bridges that still function today, and Charles Babbage built his first *thinking machine*. The utility of the slide rule continued in New York with soaring skyscrapers. As the ultimate accolade, Buzz Aldrin took an eleven-dollar slide rule to help on the first moon landing aboard *Apollo 11*.

Charles Babbage, a professor of mathematics at Cambridge University, used logs and trigonometric tables for his work. But, like all compiled data, unfound hidden errors troubled Babbage with undiscovered mistakes. Babbage was so frustrated he was motivated to find a flawless alternative that removed all mistakes. He went as far as to print the results as table data directly from his first *thinking machine* to avoid typesetting and all other printing errors. By preventing every source of error, he created the first flawless computer, the Difference Engine[81].

During the Industrial Revolution, Babbage invented two flawless mechanical engines. His Difference engine was a pioneering prototype; it only resolved polynomial expressions, while his unfinished Analytical Engine[82] supported mathematically programmed algorithms of any functionality. Ada Lovelace explains how to do this as a scientific programming language for the *thinking machine* in her translation from Luigi Menabrea's 1842 *Sketch of the Analytical Engine Invented by Charles Babbage Esq.*[83] Ada's example, once functionally tested, is eternally flawless. It is not only the first program but will forever be the longest-lasting one. After almost two hundred years, mathematics remains true to its purpose without needing patches.

In her attached notes, Ada explained how functional, symbolic programming will change the world, how computers can compose music and art, and the full scope of scientific media. She needed no operating system or compiler because the functions of Babbage's *thinking machine* were scientific, flawless, and provable. It is a remarkable demonstration of trust in computer science. She programmed the *thinking machine* using pure mathematics, the same symbolic language written on whiteboards at school and university. The language of symbolic mathematics simplifies programming and lowers costs by reducing the time to find and correct errors. Plessey Telecommunications appreciated this advantage when implementing the Capability-based PP250. The small team of newcomers used the built-in IDE hardware to enforce programming best practices, detect bugs while writing the code, detect malware, block attacks, and, most importantly, avoid patching delivered code. Scientifically defined object-oriented machine code lasts far longer without costly updates.

Ada's mathematical procedure is unchanged to this day. She never had the chance to debug her program, but once tested, it would

---

[81] The Difference Engine
[82] The Analytical Engine
[83] Sketch of the analytical engine.

pass every test, including time. Consider the impact of software with a half-life to match her Bernoulli program of approaching a hundred years. Ada's notes show that her mathematical and symbolic machine code is correct and readable today without any compiler or Integrated Development Environment. All she needed was a chalkboard or a paper pad. Her example program only needed twenty-five precise mathematical machine instructions to calculate the numbers of the Bernoulli series. Her self-standing algorithm, encapsulated as a λ-calculus function abstraction, is a perfect example of engineered software. It is functionally stable as an ageless functional algorithm packaged as a class networked by the λ-calculus. Any algorithm correctly expressed mathematically never changes or needs a patch, a compiler, or an operating system. These mathematical programs, like the abacus and Ada's program as a function abstraction, are demonstratively proven to last forever.

Now, one can see the total value of Alonzo Church's solution to the Babbage Conundrum. It is the science of flawless programs. The software lasts forever once tested as an engineered, encapsulated, type-safe implementation like the algorithms on a slide rule. Like the mechanics of the abacus. Like Ada's Bernoulli abstraction. Including the software for the *Dream Machine*. Programs survive, bug-free and safe for eternity. Upgradable from machine to machine without freezing the hardware and without the additional costs of unneeded baggage.

How different is this from the never-ending effort spent patching compiled binary programs in overstretched computers? Add the extra expense of the overheads of the centralized Black Hole. The accumulated proprietary baggage reinvented for over half a century by the dictatorial suppliers must end before it ends democratic society.

Babbage once designed his mechanical clockwork of the *thinking machine* to perform mathematical functions. Likewise, the digital ironworks of a clockwork capability-based PP250 is another example that stands every test of time. These computers democratically replace von Neumann's centralized, dictatorial mistakes with private, individual threads of cellular computation. A namespace lists the authorized and named cellular objects as named functions and data. As taught worldwide, mathematics should teach students computer science as they study and learn. Babbage and Lovelace showed how this works as they demonstrated flawlessly democratized, complex, programmable computer science that lasts indefinitely. Pure mathematics is the complete solution to the Babbage Conundrum.

Every opaque complexity of binary cyberspace disappears when exact, flawless results are calculated dynamically by perfect scientific machine code using the clockworks of the λ-calculus. Babbage adapted Napier's bones into rolls without length limitations or rectangular constraints. His endlessly rotating cogs, wheels, and drums used studs, like a music box, programmed for scientific functions instead of music. Algorithms that calculate and print mathematical science yield flawless results mechanically, avoiding every interfering error.

He implemented this idea as a working prototype, even generating tabular results for direct printing to avoid printing errors and reduce checking for hidden errors. He proved it to Ada Lovelace, and she dubbed it the *thinking machine*. Babbage envisioned programmed algorithms driven by steam. Lovelace used his infinitely capable replacement of the slide rule to flawlessly resolve programmed mathematics to fourteen or more significant digits, democratically extending the power of individuals and society through the science of mathematics.

Programmed mathematically and executed mechanically, Babbage's *thinking machine* avoids the physical limitations of the earlier rods and rules. But even more importantly, all human errors, mistakes, and corruption disappear. The *thinking machine* achieves flawless results without mindlessly trusting tables or an obscure programming language. By 1823, he confronted and solved the problem of undetected, hidden errors, slips from best practices, and mistakes from wear and tear, mechanically and flawlessly.

His publication won the Royal Astronomical Society's first gold medal.[84] Alas, his work was unfinished. Like other inventive minds, he fell out first with his team, then his backers, because his mind kept moving on, and the biggest threat became a better idea. His prototype was ornate and intricate, but no obstacles remained to his complete machines. It was an underrecognized but singular moment

---

[84] "On the Theoretical Principles of the Machinery for Calculating Tables," Online Encyclopedia Britannica Inc., 2012, web, July 7, 2012, http://www. britannica.com/EBchecked/topic/725541/On-the-Theoretical-Principles-of-the-Machinery-for-Calculating-Tables.

in the history of computer science, not only for the hardware but, significantly, for the symbolic, scientific, functional programming machine language for reliable software.

Babbage solved his conundrum through accuracy and reliability, the root cause of all issues with the binary computer. These problems cause cybercrime from centralized sharing and blind trust in programming best practices. The right solution is always the same. It is the flawless cross-checking performed on the spot by hardware to guarantee pure mathematical calculations. Physically encapsulated individual computations are confidential threads kept data-tight and function-tight as algorithms. The algorithms link together using functional tokens to form fail-safe computational expressions for concurrently threaded yet private calculations.

Only one small, encapsulated binary computation executes at a time. One algorithm in one computational process as one private instance, in a personal thread as a dynamic, symbolic, cellular computation, producing a flawless result on demand. The simplicity and purity of the mechanics are the two principles of the λ-calculus using object-oriented machine code as a clockwork computer. It is how the failsafe, object-oriented machine code of PP250 achieved impressive decades of long-term, engineered software reliability and system stability.[85] Building complex computer solutions for the everlasting future is not a question of designing ever more potent monolithic compilations with ever more complex operating systems and bigger and bigger compiled and shared images. It is the opposite, developing simple algorithms that fully meet the modular requirements of the Church-Turing thesis as functional machine code.

What was true scientifically for Babbage remains true today. Compiled data, printed as tables, hides corruption, like the binary pages of virtual memory. Interference and wear and tear must be detected. The internet works nonstop, and a cold restart of a mainframe and a new shift no longer work. Digital corruption grew from being out of the question for von Neumann to being out of control today and the root cause of cybercrime for unprotected binary computers.

---

[85] Civilizing Cyberspace: The Fight for Digital Democracy, K J Hamer-Hodges.

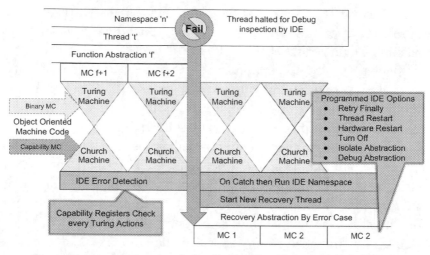

Figure 9. Automated try-catch-finally Machine Code on Fail-Safe Error

The problems are the same as Charles Babbage faced. It is the Babbage Conundrum. He revealed the pieces that solved his puzzle. Functional, symbolic mathematics over static sharing, dynamic binding over static binding, private data over public memory, guarded scope over global interference, and engineered simplicity over opaque complexity. The choice is between two classical alternatives.

- A binary computer that mindlessly trusts compiled centralized procedures sharing the same address space. But it is fragile. Corruptible pointers can damage all four corners of shared, centralized memory. The systemic weaknesses tolerate bad practices, flaws, malware, and undetected uncontrolled errors.

- A networked distributed alternative that dynamically limits results to the context of cellular instance data in a private thread of atomic function abstractions. The scope is protected atomically, instance by instance, class by class, object by object, and life cycle by life cycle. Solid engineering using measured failure rates. Engineered component boundaries and digital ownership, using typed and guarded borders to detect and prevent errors and bad practices.

The only way to protect cyberspace is symbolically, object by object, instance by instance, class by class, thread by thread, and namespace by namespace. Building cyberspace to follow the laws of the λ-calculus automatically prevents cybercrime, simplifies the design, uniformly scales the details to be equal to all, and simultaneously prevents dictatorship and cybercrime.

Unavoidable rules of ownership and individual authority are atomically enforced across time and space using the digital ironwork of Capability-based addressing as a neutral, independent hardware mechanism to enforce the laws of the λ-calculus. Using science is without bias, human best practices, outside interference, or supplier prejudice. This impartial mechanism is a second factor, providing scientific guard rails governed by the laws of nature, not by humanity. The guard rails run non-stop as a uniform check on every step of the atomic and cellular information boundaries from edge to edge of cyberspace. Regulating access rights through proof of authority detects all mistakes through the automated, neutral, type-safe, and fail-safe mechanisms of mathematics and logic.

As proven by PP250, no monolithic, centralized operating system or privileged hardware modes are required. They are removed with all the other baggage of branded binary computers, starting with the hidden superuser hardware and dangerous machine instructions of page-based virtual memory. Avoiding these branded inventions, using the λ-calculus requires no human workarounds to achieve flawless scientific results. Only the scientific function abstractions as an application namespace of cellular atoms of dynamic digital functionality are needed. They create a living digital species that operates as an independent software robot, even AI-enabled. Computer software is future-safe when written as mathematical and logical machine code. The software becomes scientific statements of truth that last forever without patching or compilers. Perfect science is vital for a digitally dependent future, and simplification is essential for the endless commitment of freedom-loving nations to life in cyberspace.

Static binding, used by von Neumann, is a dangerous, easily corrupted mechanism effortlessly abused. At the same time, dynamic binding avoids ambient authority by limiting power to the atomic functional modularity of the λ-calculus. Constraining programs to these scientific guardrails create object-oriented machine code in readable application-oriented form.

Every digital detail is an individual instance of a parent class, a protected atom of digital cyberspace. If errors exist, they are found on the spot. No single points of failure exist. The hierarchal, need-to-know namespace forms a detailed security matrix of named, immutable links. The digital ironworks of Capability-based addressing bind links dynamically into a fail-safe, scientifically programmed, hardened digital machine. The program-controlled Church machine works in harness with a Turing machine, realized side-by-side as the cockpit of threaded computations. The underlying binary data computations are hidden and protected from abuse, accidental interference, and undetected attacks.

This method of dynamic binding prevents unauthorized access. Only key owners can open any gate, hiding all internal details from attack. Furthermore, any top security class can insist on further checks using another immutable key as a security token, a password, or a user identity. It provides two-factor security that protects the underlying instance of a function abstraction from externally crafted attacks. These keys to the digital ironwork of Capability-based addressing are immutable, unlike binary pointers in virtual memory, constantly exposed to fraud and forgery.

Given the tragic consequences of flawed computer science, a computer-dependent AI society will not survive in peace and harmony while using binary computers. AI-enabled cybercrime and cyberwars threaten life in a growing variety of traumatic ways. When the nation plays with fire, everyone gets burned. It is an untenable situation. It is a threat to public order and social stability. The only option is to remove the malignant cause by regulating the computer industry as a public utility and protecting the rights of innocent civilians.

The Babbage Conundrum impacts not only individuals but also democracy. If not resolved for good, this riddle of computer science will lead to Big Brother or a digital inferno. In either case, democracy burns to the ground. Industrial dictators and governments will resist unless and until the public demands action, or we start again after the ultimate chaos of global cyber catastrophes.

POWER CORRUPTS

The Danger of AI and Centralization

In a world where artificial intelligence melds with dangerously outdated binary computers, a looming catastrophe of corrupt centralized power threatens to dismantle the foundations of a democratic cyber society. AI makes everything worse. When digital technology was expensive, John von Neumann shared memory in the binary computer. His first-generation binary computer used primitive linear procedures, fully exposed in shared memory.

At first, sharing was minimal until the imperial superuser model of Multics and authoritarian time-shared virtual memory took off. Then centralization and sharing became the dangerous blueprint for every subsequent decade. Centralized sharing is foreign to nature. The digital gaps, cracks, and voids that empower malware emerged by discarding science. However, the centralized powers of the operating system, refined for over half a century, have exceeded their utility. Costs still escalate, and projects still fail as staff shortages mount. The dictatorial architecture is opaque, cumbersome, unstable, and dangerous in the age of AI-enabled global cyberspace.

The binary computer updated by Multics time-sharing for mainframes is a centralized dictator, a superuser authoritarian that hides the dangerous digital mechanics of virtual memory. It caters to the privileged and those who dominate computer science as empire-building dictators. At the height of the Cold War, the allure of dictatorial digital supremacy, bolstered by Big Brother surveillance techniques and other digital technologies, captivated the mindset of that day.

Amidst this backdrop, binary computers took the path of centralized dictatorship to manage enslaved users and ignore the science from 1936. Even in 1972, when PP250 disclosed a better alternative, industry, and computer scientists clung to John von Neumann's shared binary architecture. It sidelined interest in the Church-Turing thesis and the efficiency of the λ-calculus symbols achieved by the PP250.

Von Neumann's ego-driven pursuit of notoriety pushed his architecture to the forefront, overshadowing the science and others' work on the EDVAC project. Blindly, the world followed, unthinkingly accepting his pioneering work as the complete and satisfactory expression of computer science. Sadly, he only addressed one-half of the science. He omitted the dynamic, functional modularity of the λ-calculus, the essential, top-down, scientific half of the Church-Turing thesis that encapsulates and hides every dirty implementation detail. The λ-calculus distributes computational functionality evenly as needed, case by case, to individual users as the symbolic tokens to function abstractions.

The allure of shared, centralized computer science swayed minds on both sides of the Atlantic. With evident flaws hidden, the designers were delighted to play God in a centralized dictatorship run by a branded operating system.

Solving the riddle composed by Church and Turing was left to those who first confronted the challenges of global applications using computer communications. The ARPANET[86] also missed the point and approached cyberspace as a physical switch between standalone mainframes instead of a logical extension of computations as networked, abstract, functional applications.

Physical addressing is computationally flawed. Not only do binary computers allow undetected malware to corrupt the virtual machine and the central operating system, but they also attack the network using Denial of Service attacks, clickjacking attacks, and cross-site attacks. All these static, physical problems disappear when

---

[86] The Advanced Research Projects Agency Network (ARPANET) was the first wide-area packet-switched network with distributed control and one of the first computer networks to implement the TCP/IP protocol suite. Both technologies became the technical foundation of the Internet as a physical rather than a logical switching system.

the λ-calculus hides the details of network endpoints. The function abstractions of the λ-calculus logically hide this information, but interconnecting computers through physical addressing created the patchworked cyberspace where malware thrives. By exposing the network's physical endpoints, warring kingdoms and criminals fight for control over individual computers, fostering undetected attacks and ignoring the λ-calculus abstraction as distributed concurrent, functional applications.

The rise of the mainframe demanded innovative scaling by the late 1960s when airlines and telecommunication entities sought computer science to enhance their global networks. Jack Dennis and Earl van Horn proposed a new hardware technology called Capability-based addressing to protect memory and abstract physical objects symbolically. The PP250 adopted Capabilities to go beyond secure memory segmentation and implement a networked multiprocessor that exploits the λ-calculus. Doing so thoroughly adds programmatic control over networked computations, concurrency, and performance distribution.

For the hardware industry, the science of the Church-Turing thesis remains unrecognized. The centralized memory of a batch-processing mainframe, the paged-based hardware of virtual memory, and the privileged operating system dominate the industry. But semiconductors and the internet changed everything. Low-cost shared memory, and networking ascended to prominence, fully revealing the threats of hacking, malware, and corruption in the centralized binary computer. For over half a century, ignoring science has only increased the lost opportunities. Unfortunately, the patchwork approach to computer communications is again like shared linear addressing, inside out. When implemented by Capability-based addressing, the implementation's privacy and security highlights the critical lesson of computer science. Fundamentally, functional abstractions encapsulate and hide all physical implementation details from all outside threats by preventing undetected attacks.

Security only exists in one way in a centralized mainframe when users log into the centralized operating system running as a superuser that enslaves clients all limited to their user identity. However, on the network side, there is no user identity for login. The binary code is a wide-open backdoor of shared virtual memory.

Downloading code includes malware, deliberately or by mistake. Through email attachments, text messages, or as a hidden activity when clicking a browser, the damage is instantaneous.

While Alonzo Church's visionary contributions linger in obscurity, binary computers evolved into a patchwork of branded stand-alone computers. Cyberspace, built from branded computers, sprawls into fragmented digital realms run by incompatible industrial dictatorships. They are ruled by suppliers and corporate users but undermined by malware, criminals, and enemies. The rise of ego-driven designs continues to champion physical over logical addressing. The industry sustains von Neumann's approach, fortified by Multics and figures like Butler Lampson. Butler, more than others, cast aside the better alternative and the promising innovation of Jack Dennis's Capability-based addressing from the start. Capability-based addressing is superior because it obeys the lessons of the Church-Turing thesis, logically abstracting functional details using uniquely named tokens instead of shared physical address space. As the alternative to shared memory, Jack Dennis's Capability idea to protect the programmed functions stored in memory segments opposed Butler Lampson and his followers.

Notably, at SOSP-6 in 1977, Butler Lampson dismissed Capability-based addressing as *"special-purpose"* for the Telecommunication Industry, where *"nothing good is ever found."* Butler asserted that lightweight operating systems were superior for detecting software errors. However, history's judgment has exposed the fallacy of his optimistic claims. Without accounting for the overheads and vulnerabilities of virtual memory and centralization, Butler's benchmarks ignored the ultimate need for compiler overheads to perform boundary checks added as in-line code overheads. The fact remains that centralized operating systems cannot fully protect themselves. Over time, the myth of centralized operating systems outperforming Capability-based addressing has evaporated,[87] debunked by the never-ending problems guarding centralized power and the ascendance of ransomware.

[87] Capability Myths Demolished, Mark S. Miller Combex, Inc. Ka-Ping Yee University of California, Berkeley, Jonathan Shapiro Johns Hopkins University

Successful ransomware attacks are the ultimate proof of the inherent danger of AI-enabled malware in patchworked cyberspace. Adopting Capability-based addressing at its inception would have accelerated progress, unburdened by the drag of ever-escalating cybercrimes. Add to this the development of unnecessary privileged hardware and complex software baggage blocking end-to-end cyberspace.

Marked by repeated failure, the computer industry's pursuit of malware is futile, self-evident from a steady stream of successful ransomware and zero-day attacks. The shared binary exposure of centralized computer powers fails to detect and prevent cyber-attacks, leaving application programs and subsystems vulnerable to undetected exploitation. The Babbage Conundrum escalated from human mistakes to systemic malware. Adding more code overheads to monitor runtime flaws for case-by-case patching is expensive, time-consuming, and inadequate. The Confused Deputy and Denial of Service attacks prove beyond resolution, leaving unresolved dangers of the piecemeal solution created by the centralized physical power of the superuser.

Compelling, detailed computer security is the only solution acceptable to AI-enabled cyber society in a world of constant threats. Capability-based addressing, guided by immutable tokens, ensures the security of networked trade routes throughout global cyberspace and the equality of privileges throughout computer science. Moreover, the approach safeguards against Ransomware, Confused-Deputy, Denial of Service, zero-day attacks, design bugs, and any malware plaguing binary computers. Otherwise, the problems of binary computers are insurmountable. Applying intricate, expensive, code-based, case-by-case patching distracts from progress, diverting the most skilled staff to fix a clunky, outdated code base. Meanwhile, the PP250 results promise a quantum leap in performance, efficiency, functionality, and future-safe scientific digital security.

The journey for safe computer software already adds improved compilers with integrated development environments (IDEs) that strive to ensure secure, efficient code images for use by virtual machines. Despite substantial progress, testing remains a formidable challenge, consuming over half the development costs of complex binary applications. But another paradigm shift beckons when adopting Church-Turing machines. Empowered by the λ-calculus, Capability-based addressing offers fail-safe object-oriented machine code with a built-in IDE. The IDE works non-stop, from the first line of code development through integration testing and, most important of all, throughout service life. Immunizing user code and distributed, concurrent complex applications from mistakes and attacks. Unlike binary computers, where compilers battle malware with code patches and unfathomable best practices, Church-Turing computers like PP250 boast inherent and unbeatable systemic advantages.

The Church-Turing computer includes a complete, inline Capability-based Integrated Development Environment (IDE). The fail-safe machine instructions pinpoint every program bug, and the built-in, readable machine code of each application-oriented function of every namespace token is quickly confirmed. The tokenized machine code is the harness for functional λ-calculus. Automatically, Capability-based addressing supports and tests computational concurrency, thread synchronization, and distributed performance locally as fail-safe machines. It is a distributed multiprogramming solution for networked, fail-safe computations across global cyberspace.

But Alonzo Church's masterpiece on computability stands idle while human egos drive the followers of von Neumann, Multics, and Butler Lampson.[88] Not only is the symbolic λ-calculus ignored, but the complimentary work of Jack Dennis and Earl van Horn, who proposed Capability-based addressing to protect shared memory also using symbols. Butler loudly led the computer industry charge against PP250 and Capabilities at SOSP-6 in 1977. Butler dismissed the alternative as a *"special case"* only for telecommunications. He claimed operating systems were lighter, better, and faster at detecting software errors. Their centralized role made this easy, he continued. History proves him wrong on every count. Capability-based addressing started when pioneering microprocessors. Computer science would have advanced faster and further together as functional microprocessors without the dead weight of operating systems, virtual memory, and cybercrimes.

---

[88] <u>Butler Lampson, MIT News.</u> the loudest supporter of centralized operating systems

Capability-based computers always win because machine code is more potent without needing the unruly powers of a centralized operating system. The computational threads execute privately, the instructions are fail-safe, and immediate testing with on-the-spot atomic corrections pushes productivity to the maximum. All this gain is because hardware checks are performed in parallel for every function and do not depend on any offline compiler or an opaque monolithic operating system. Over the long haul of history, the neutrality of the mechanics will prove invaluable for cyber democracy.

The goal of the operating system catching malware failed resoundingly, repeatedly demonstrated to this day by successful ransomware and new zero-day attacks. But namespace tokens lock malware out of the cockpit. Preventing binary computers' other unsolved problems, like the confused deputy and denial of service attacks, is critical. Functions hide the dirty details while clarifying every trade route in cyberspace. The endless complexity of case-by-case patching evaporates.

Furthermore, testing represents over half of the effort and cost of software development. Integrated development environments (IDEs) simplify testing and allow AI to write more code daily. This code comes with new threats of hidden malware. But a Church-Turing machine builds readable, secure, fail-safe machine code, a perfect IDE framework to test and debug programs online. A built-in Capability-based IDE is a significant win for a *Dream Machine,* the fastest and most economical method for fail-safe, long-term software design and development.

Moreover, functional programs using object-oriented software machines and Capability tokens resolved by a namespace significantly improve programmed concurrency, synchronization, distributed performance, and flexibility. Like JavaScript today, Capability tokens were *'first-class citizens'*[89] long ago. PP250 passed Capability tokens as arguments to other functions and returned them from functions using the object-oriented machine code. The power of symbols exists uniformly throughout networked cyberspace, unhindered

---

[89] In a given programming language design, a first-class citizen[a] is an entity which supports all the operations generally available to other entities. These operations typically include being passed as an argument, returned from a function, and assigned to a variable. https://en.wikipedia.org/wiki/First-class_citizen

by primitive operating systems, resolving high-order expressions end-to-end across cyberspace without a compiler as object-oriented machine code.

The ultimate advantages remain unappreciated by the computer industry. However, the software becomes a hardened component with a measured reliability to engineer cyberspace. The MTBF of software that drove Plessey Telecommunication in the 1970s has returned. The flawed science of monolithic software has resurged to undermine democracy and democratic nations. Solving the Babbage Conundrum requires scientific boundaries expressed by the two sides of the Church-Turing thesis.

Wars in cyberspace accelerate the development of increasingly dangerous weapons. 'Pay-or-leak'[90] ransomware now demands Bitcoin payments on short notice. Gangs in Russia, China, Iran, and North Korea have already deployed pay-or-leak ransomware. Centralized AI cybercrime is as easy as turning on a light. Nothing is safe, and nothing in a binary computer can prevent this. It is only a matter of time before a global catastrophe occurs, and AI will only accelerate the speed and scope of the impending disasters.

Only flawless computer science is an acceptable digital extension for humankind. Backward compatibility must end if a computerized society is to progress and flourish into an endless future with AI-enabled organization improvements. Mashing binary computers together using unscientific, branded platforms only increases the centralization of power and Orwellian results. It is life or death for industrial dictators to prevent a coup d'état attack in cyberspace, as in ancient Rome. Ransomware and zero-day attacks are the Ides of March.[91] These attacks succeed because they cannot block the sins[92] of human behavior. Thus, the attacks will continue until we replace von Neumann's centralized shared architecture.

---

[90] CrowdStrike 2023 Global Threat Report
[91] The Ides of March is the seventy-fourth day in the Roman calendar, corresponding to March 15. It was marked by religious observances and was notable in Rome as a deadline for settling debts. In 44 BC, it became notorious as the date of the assassination of Julius Caesar, which made the Ides of March a turning point in Roman history.
[92] According to the standard list, they are pride, greed, wrath, envy, lust, gluttony, and sloth, which are contrary to the four cardinal virtues of prudence, justice, temperance, and fortitude, plus three theological virtues of faith, hope, and charity.

Winston Churchill considered democracy the least imperfect of all alternative forms of government. However, the concentration of centralized binary computer powers corrupts society in ways never before experienced in the history of civilization. Orwellian dictators wait in the wings as cyberspace grows ever more dangerous, weakening citizens, and as learned from Marshall McLuhan, the medium is the message, and the message is corrupt. Crime always wins.

Instead of centralizing power, cyberspace must be democratic, distributing authority scientifically through symbols equally and evenly without superuser hardware or a privileged operating system. Individual names as functional symbols speak the language of the λ-calculus. It allows science, mathematics, and networking to distribute power and logic equally. Equality, freedom, and justice lead to happy personal fulfillment, with the certainty of science based on individuals exercising power incrementally. It is the ideal case of AI-enabled empowerment.

Mathematics is not only equal for all, understood, readable, and applicable to any subject worldwide; it also survives. Mathematical machines also survive because the machine language never changes. For example, the abacus performs the calculations:

$$a + b = c \ or \ a - b = c$$

The expressions never change, and the flaws of the ego disappear. The Red Queen has no foothold. Science drove the abacus to survive forever. It will always be so, leading the way for a Dream Machine, running fail-safe software that passes every test of time. Programs that endure while hardware improves over centuries, as profoundly demonstrated by Ada Lovelace.

To keep the dictator's grip on evolving threats, industrial suppliers increase their unfair privileges, blocking progress while failing to stop attacks, simultaneously hurting users in a downward spiral. A face-off between the suppliers, software dictators, and enemies fighting each other to dominate binary cyberspace. It is an international war of escalation to capture keystrokes, data, credentials, and mouse clicks. As spies and gangs successfully perform attacks worldwide, the only option is to add controls that further subjugate the citizen users.

Every source of power worldwide, from governments to businesses, crooks to dictators, individuals to political parties, and hardware manufacturers to software suppliers, take part if they can gain an advantage. Nothing stops them, and software has no conscience or ethics. AI makes everything worse. The laws of the land are ignored and bypassed. Only the λ-calculus can reverse this existential, terminal condition and deliver actual computer science with the raw, universal power of magical mathematics and the efficiency of λ-calculus reduction.[93]

Centralized, dictatorial cyberspace tilts against innocent citizens, favoring the dictator, criminals, the rich, and the powerful. Governments, industry, spies, gangs, and enemy armies redesign old attacks, invent new crimes, and discover design flaws to use as the next big zero-day catastrophe. Aided by the superhuman powers of AI software, crimes will grow in number, turning individuals against friends and neighbors. The suppression of information and honest debate is not limited to Russia, China, Iran, and North Korea but also occurs in the USA. [94] And not just elections are under attack. The dictatorial power in binary computers corrupts civilization and democracy locally and from the earth's far corners. It destroys the progress of society. Already, industrial dictators have too much control; their Orwellian endgame is frighteningly close.

Invisible software with superhuman powers of corruption and instantaneous global reach is a new extension, but not of humanity. AI-enabled dictators will crush civilization. Misdirected AI software will manipulate the digital details, making recovery impossible because patching gets harder. Artificial intelligence, deep fakes, rewriting history, and creating the unperson help the dictatorial binary computers to advance a return to Orwellian *1984*. As AI improves and expands in cyberspace, the Red Queen undermines law and order in society. The growth of industrial tyranny is unstoppable. Individuals suffer, society malfunctions, and the progress of civilization moves into reverse. Everything about binary

---

[93] Untyped lambda calculus does not distinguish between different types of data. This means a function can operate on numbers or any other type, such as strings, and any other non-number object found in nature or abstracted in virtual reality.
[94] Twitter suppression of Hunter Biden news.

cyberspace is an existential, global threat to individuals, businesses, organizations, and nations.

The computer industry shows no interest in adopting the dream of perfection the λ-calculus offers. The branded solutions may be outdated, but as industrial dictators, they dominate the market. They always dismiss ideas that disrupt their business. But science must win for computers, software, individuals, and society to progress. Science is the only way to go. Removing the centralized power of the superuser and the operating system turns cyberspace into science, converting computers into the *Dream Machine* designed for a globally fair Public Utility that is scientifically secure, unbiased to everyone worldwide, more productive for all, and longer lasting.

Digital security is imperative. Enforced by the guard rails of Capability-based addressing, the advantages include functional programming in machine code that extends across cyberspace without interruptions caused by branded operating systems. Well-named tokens make data structures and programs transparent as digital abstractions. Programmed functions without unintended side effects or malware entering the cockpit of computer science. This decentralized, distributed, concurrent paradigm shift secures the computational structure to democratic, private extensions of individuals. Computer users will then pave the way to a survivable digital democracy. The immutable digital tokens of Capability addressing elevate machine code programming to a readable, functional programming language. It is modular, secure, private, and efficient, without the baggage created for binary computers.

Binary computers, once hailed as the vanguards of progress, now stand as antiquated relics of the past. Their centralized, dictatorial architectures stifle democratic values while allowing undetected cybercrime to flourish. It is the centralization of power that corrupts the shared binary computer. As the world grapples with escalating cyber threats, binary computers and AI programming become an explosive hazard—a WMD with unchecked AI power for malware creation and disinformation. AI campaigns erode the very fabric of society. The centralized powers that rule binary cyberspace stifle individual freedom and democratic progress, pushing civilization over the precipice into Orwellian digital dictatorships.

The computer industry's complicity in sustaining these flawed systems is evident, prioritizing profit over progress. The need for change looms ever larger. It is hypocritical to use binary computers for computer science. The computer hardware industry cannot ignore the λ-calculus. As the foundation of all software theory, λ-calculus makes the future of AI-enabled nations safe, moving them forward the way they need to work. Science is the antidote to every existential cyber threat.

The industry must align with Babbage's timeless truth on mathematical logic. Cast aside industrial dictators and embrace the future, not the past. Civilized society builds on equitable, democratic principles where individuals flourish. The path forward demands this equity. Only the science of the Church-Turing machine has the elegance of λ-calculus. Symbols are the scientific secret sauce. The immutable tokens of Capability-based addressing, the rugged mechanics of λ-calculus, and the efficiency of object-oriented software all work this way.

Transparent distribution replaces opaque centralization, and individuality and simplicity trade for malware and corruption in this transformative evolution of digital cyberspace. The mathematics that underpins computer science, as found throughout nature, is inherently democratic and understood scientifically across borders and disciplines. A computer without a stranglehold on power. Without undetected malware. Computer science in the form that channels the principle of equality and individual empowerment. The language of mathematics as machine code underpins this endeavor, guiding the way to a society marked by harmony, equality, software that survives the tests of time, and unfettered democratic progress.

Yet, the battle rages on. The threats of binary computers fortified by AI are the breeding ground for increased dictatorial power and catastrophic cybercrime, propagating discord and fear and increasing civil conflict. The struggle between industrial dictators and citizens intensifies as centralized power systems accumulate, thwarting progress and crushing democratic values. The addition of AI amplifies the dichotomy between centralized control and equitable progress. It is a critical juncture forcing humanity to decide its digital destiny. The scientific path illuminated by Alonzo Church is calling like Henry Briggs ever louder. Like John Napier, we must walk away from decades of hard work for equitable, civilized progress. Alonzo showed how to avoid the looming threats, not only the threats of malware and corruption but also of AI breakout. Things will be worse until actual computer science is the essence of digital civilization, as it began at the birth of society with the abacus.

INDIVIDUAL FREEDOM

Realizing the λ-Calculus

The power of mechanical abstraction is not new. The abacus and the slide rule use numeric abstractions. Babbage's *thinking machine* allowed Ada's Bernoulli program to use function abstraction. It uses this far more powerful programming paradigm where mathematical functions are variables. When Church and Turing studied the *Entscheidungsproblem,* they approached their solutions to computability from these two alternative directions. The Turing machine worked bottom-up using binary hardware to abstract numbers used in programmed algorithms.

In contrast, Church worked top-down using the language of λ-calculus to manipulate function abstractions. But they work together, as demonstrated by the PP250, but more importantly, as proposed for a *Dream Machine,* the perfect computer for global cyberspace. Alan Turing's bottom-up solution runs one algorithm at a time, sufficient to ground one isolated term in an atomic computation thread, becoming the λ of the λ-calculus.

Functional programming improves programs by avoiding state conditions and the mutable data of a binary computer. Symbols represent expressions passed as arguments, returned as results, or assigned as variables in a declarative style that focuses on results instead of choosing the imperative steps to get there. PP250 experience shows the power and stability achieved using Capability addressing as what programming now calls first-class citizens. One by one, the symbols translate into real-world digital objects processed by λ-calculus threads. The immutable tokens of Capability-based addressing work as machine data types, stored as variables, passed as arguments, returned as results, stored in structures, created on-demand, and destroyed when finished. Hence, the machine code supports first-class functions and objects.

The *Dream Machine* is a computer that integrates Turing's idea on binary computers with Church's ideas on the λ-calculus. Alonzo's symbols representing functional objects become a second machine type of immutable Capability tokens. A Turing machine manages the original binary type and runs the selected binary machine-coded programs. A Church machine handles the six λ-calculus instructions. Each λ-calculus symbol is an immutable token referencing a Capability limited, digital black box as an offset in the namespace table. For example, each thread of computation is an instance of a hardware-accelerated thread abstraction. Real-world events initiate these active, individual, and personal calculations. A reserved Capability register lists the object details of the namespace, and another the active thread, as a data structure for the LIFO stack and storage for each data register and each Capability token belonging to one private computation strand. The set of Capability registers limits the scope of machine computation characterized by one private thread in one specific namespace.

The dynamic threads of computation used the services of other function abstractions accessed as protected subroutine calls. Each call to an *Enter-Only* Capability token saves the token on the stack as a breadcrumb trail. Thus, computer science as a Church-Turing machine hides the details from the top. However, exposeing the binary details atomically at the bottom as, when, and where needed in each private thread. Each unique λ-calculus symbol is a Capability token, a key to unlock access to a digital object using a Church instruction and the namespace table. The λ-calculus names include function abstractions that express the nodes of the application as a coherent hierarchal namespace. The namespace excludes unapproved digital objects, thus preventing outside interference and any malware from entering the cockpit of a *Dream Machine* as a computer. The cockpit is Turing's simple binary machine, updated to execute both Church and Turing instructions in harness side-by-side.

An application is a set of approved, immutable Capability tokens structured as a hierarchy of keys to unlock the programmed digital objects in a defined order. The internal details remain hidden from outsiders, only accessible to the cellular step of the active thread. As a node in the namespace hierarchy, the protected atomic instance is a function-tight, data-tight, type-safe black box. Consequently, any attached hardware, a keyboard, a medical device, or another remote function abstraction is linked symbolically and approved by the namespace.

Six Church instructions link the λ-calculus symbols to the computer hardware registers using the namespace table to map the top-down symbolic names to the bottom-up tokens. The Capability registers define the boundary limits of the thread as a cellular computation of function abstractions. Threads weave a private course of atomic subroutine calls to functions linked as a hierarchy of nodes and terms in symbolic expression. These personal computations work together locally and remotely, proceeding concurrently and communicating privately to synchronize activity across a global network. The details of the Turing machine, the memory system, and the web of connectivity are all hidden. Dynamically bound together by the Church instructions following the namespace tokens and laws of a hierarchical namespace.

The Turing machine, as a simple binary computer, needs no virtual memory or a centralized operating system. As the engine of a functional λ-calculus namespace, there is no branded hardware, superuser privilege, or centralized activities. Only immutable tokens represent function abstractions and basic digital objects. Every hardware instruction is available for use and crosschecked by Capability tokens. No hidden instructions exist on a level playing field using immutable tokens as keys to unlock defined functions in a need-to-know hierarchy. Each functional node has a private list (a C-List of Capability tokens) that structures the application's need-to-know organization. The tokens regulate power incrementally through the application's software structure. So, for example, the centralized monolithic operating system is decomposed and distributed as named function abstractions. Structured as object-oriented classes, storage management, thread management, and input-output management become fail-safe services.

The programmers decide on symbol names and code the machine code classes, specifying the access limitations to each named object.

PP250 limited access to *Read, Write,* and *Execute* for data objects and *Load, Store,* and *Enter* restrictions for Capability tokens. So, for example, the abstractions for scheduling, store management, and threads are tokens permitting *Enter*-only access, and the digital content remains hidden, only exposing functions by their symbolic name. An *Enter* token identifies the nodal *C-List* to a λ-calculus function abstraction to call an internal named function in the abstraction's *C-List*. These tokens permit computation access to the *Push* and *Pop Call/Return* Church Instructions linking cells of threaded functionality.

In contrast, *Execute* access allows an identified object to be opened as machine code and take over programmed control without changing the thread stack. These two access rights define the computational limits by reloading critical Capability. Equivalent to setting the paper tape of the Turing machine managed by the laws of the λ-calculus. At every step, the namespace translates the tokens identified by the thread. As a function abstraction, the namespace also provides memory and scheduling services.

The mechanisms protect the function abstractions from outside interference and internal bugs. At the same time, the hierarchy correctly transfers control to the following atomic step while keeping everything else, including the prior actions, hidden. These protected, threaded computations implement the λ-calculus across a namespace as a functional subsystem in cyberspace. The named symbols as tokens hide the binary implementation as details using Alonzo Church's masterpiece.

Like algebra, λ-calculus defines an abstract mathematical system; both are similar yet different in specific ways. Like capability-based addressing and object-oriented programming, they all use names as symbols to reason about objects in expressions using rules. Algebra manipulates symbols as quantities that can often be represented graphically by equations, terms, and graphs to model and solve engineering problems.

Alternatively, the formal system of λ-calculus is a framework to express and manipulate functions as programs. In λ-calculus, the fundamental operation applies programmed procedures to arguments. The PP250 matched the λ-calculus symbols to Capability-tokens used in the object-oriented machine code as private and secure computations. These computations enforce the laws of λ-calculus from edge to edge across cyberspace, leaving no gaps or voids that malware can exploit.

The critical difference between algebra and λ-calculus is that algebra, like binary computers, only understands symbols representing specific values or given quantities. On the other hand, as Ada Lovelace foresaw, the λ-calculus deals with functions that apply to any kind or type of input value. These abstractions enable the laws and order of democracy to be digitally realized and scientifically protected. When Ada Lovelace solved her Bernoulli abstraction, she used this feature of functional programming that allowed her to pass another expression as a variable and as she imagined any other form of abstract functionality. This enormously powerful capability is missing from binary computers and algebra.

Algebra focuses on solving equations to find specific values, while λ-calculus focuses on defining and manipulating functions. This critical difference allowed a single readable statement of PP250 machine code to express and hide the networked implementation of a private, personal, international telephone call.

myPhoneCall = TelecommunicationNamespace.Call(myPhoneBook.Mother)

This powerful, single-machine Church Call instruction returns a functional token that manages *myPhoneCall*. It needs no operating system or language compiler. Instead, it uses tokens that abstract the *TelecommunicationNamespace,* including a *Call* function and *myPhone* book abstraction with a functional entry abstracting *myMother.* The returned connection, *myPhoneCall,* is another function abstraction, this time for a phone call that might be local or could stretch worldwide. The differences between these two very different kinds of phone calls remain hidden and managed internally.

Moreover, the declarative machine code is clear to both the reader and the computer and fully appreciated by all, even amateur programmers. The best programming practices that stretch around the world are built-in and secure. Both algebra and λ-calculus are robust mathematical systems with broad applications to problem-solving. However, the functional advantages of the λ-calculus make it ideal for implementing computer science as networked cyberspace.

Because everything in the λ-calculus is symbolic, the six Church instructions use a protected namespace table to translate tokens into accessible locations that store digital information with type-limited access rights. This private, logical action replaces the physical indirection needed to implement shared virtual memory.

PP250 used Capability-based addressing to hide the implementation details of every type. It includes programs, physical memory, working data space, and attached equipment while allowing the owner private access requests. The tokens link an owner to a target, defining the machine type as Data or Capability with mode limitations. For example, an executable binary procedure might use some *read*-only and some *write*-only binary data for input and output. An *Enter*-only token represents a λ-calculus abstraction, identifying the tokens as a list that structures the implementation. The immutable tokens are computational tickets held in lists that Jack Dennis called a *C-List* in 1965. The functional mechanics of Jack's Capability-based addressing add deep and detailed guard rails, pipelined with the instructions, using dedicated Capability registers. The Capability registers implement the guardrails of the λ-calculus expressions.

As an application, the namespace tokens allow functional statements beyond pure mathematics. Ada recognized this long ago and expressed her vision when she understood Babbage's design for his unfinished masterpiece, the Analytical Engine. In a letter to her friend and mentor, she described her belief that computers served far more than Babbage's mathematical calculations, only intended for tabular numeric calculations.

Ada saw into the distant future. Her vision understood the abstract power of functional programming to extend the ability of any individual. She demonstrated this when passing functions as variables. She envisioned a world that manipulates text, sound, music, pictures, and everything we call media. Ada wrote, *"In enabling mechanism to combine together general symbols in successions of unlimited variety and extent, a uniting link is established between the operations of matter and the abstract mental processes of the most abstract branch of mathematical science."* She perfectly described the λ-calculus. Her notes A to E, attached to her translation of *Luigi F. Menabrea's "Sketch of the Analytical Engine,"* make it abundantly clear she recognized unlimited power *"to act upon other things besides numbers,"* as abstract objects from the science of operations.

She gave functional examples, including painting pictures, composing music, editing sound, and daring to communicate. She also understood the limitations, writing, *"The analytical engine has no pretensions to originate anything. It can do whatever we know how to order it to perform. It can follow analysis, but it has no power of [to] anticipate any analytical relations or truths. Its province is to assist us in making available what we are already acquainted with."* These limitations remain today even for AI-enabled society. AI is just a tool and cannot harm when correctly used. It can only jailbreak from a binary computer. AI must be constrained and targeted for a functional purpose by the symbolism of mathematics. The only way to do this requires the full measure of modular science to illuminate and eliminate corruption in cyberspace.

For PP250, machine code is declarative, and digital objects like function abstractions have names used by the machine code instead of shared physical addresses. Each Capability token uniquely defines a secure top-down digital target or a proxy to a networked target. Each target is type-limited, size-limited, and access-limited. The Capability tokens encapsulate and hide Turing's imperative bottom-up binary addresses from malware. The λ-calculus modularity prevents the threats and dangers of direct access in the binary computer's overstretched and exposed memory space. Avoiding centralized sharing simplifies computations as individual, distributed, type-safe, data-tight, functional threads. The hierarchical namespace is assembled incrementally as a structure of nodes using need-to-know names. The uniquely named function abstractions reference other nodes in an application as a hidden hierarchy of limitless objects as a secure application of approved symbolic, self-explicit terms held as a namespace hierarchy.

Using names instead of a shared linear address converts monolithic compilation into self-explanatory, readable, object-oriented machine code. The ability to write object-oriented machine code (λ-MC) dramatically simplifies the foundation problems of computer science. Malware attacks are locked out by excluding unknown objects from the namespace, preventing unauthorized programs from entering the cockpit. The names in the namespace predefine the operations in the application's readable (high-level) machine language. The dynamic translation of each named term double-checks the validity of every memory reference dynamically at run time. Virtual memory uses indirection to locate memory pages. Still, in the *Dream Machine,* this necessary indirection is used as the pivot between the top-down logic and the bottom-up implementation.

It is the fulcrum of cyberspace, a dynamic opportunity to cross-check every machine instruction and every memory reference logically and physically, which is impossible with the physical binding of binary computer instructions. The check and balance of the λ-calculus validate every instruction step. The memory hides because the laws of the λ-calculus encapsulate every binary command enforced and validated by the digital guardrails of Capability-based addressing.

The fail-safe instructions that obey the laws of the λ-calculus detect malware on the spot. Unknown software cannot invade the cockpit or the thread because actions cannot exceed the guardrails. When Ada Lovelace wrote the first program circa 1840, she used Babbage's pure mathematical machine code. As shown in her documentation, she used the same syntax found in classrooms worldwide as written on chalkboards for students to resolve. Using flawless mathematics as machine code prevents all sources of interference. Any remaining threats are correctable bugs due to poor specification or inadequate testing. Even these cannot turn into zero-day attacks because the fail-safe machinery detects the bug before causing any harm, pin-pointing errors and allowing for speedy resolution without monolithic recompilations.

Mathematics never ages. The language of science only improves. Starting in Babylon, arithmetic was future-safe, as the continuous use of the abacus proves. Likewise, the mathematics of a *Dream Machine* uses the science of the λ-calculus for future-safe, corruption-free software. The tokens firmly transfer power to the user, the citizens colonizing cyberspace. Colonizing cyberspace this way powers digital democracy that can survive in the hostile digital world of global cyberspace.

Networked cyberspace must work for individuals, unlike centralized dictators working against democracy. The machine language allows different namespace applications to support cyber colonies using new laws. Functional modularity defines a

computational DNA for each namespace defined by the atomic
hierarchy of modularity instead of monolithic indifference. Settlers
dwell side by side, each with a private namespace. All work from the
same scientifically level field of play that removes dictatorship rejects
malware, and detects mistakes on the spot.

In a cyberdemocracy, the power to lead must be democratic.
Consider, for example, how a queen bee uses her smell to prevent
egg production in her rivals before killing them. Her ability depends
on support. The bees select their queen, not by votes, but by smell, a
functional process programmed by their DNA. The algorithm exists
as an arrangement of function abstractions. Like a Supreme Court
judge, it is a lifetime appointment.

In general, democratic power is time-dependent. A mayoral chain
of office is a classic example. The chain adorned with the rights is
loaned to an elected individual to make decisions on behalf of the
electorate. The lord mayor of London is prestigious but primarily
ceremonial, picked annually, holding the title for one year. The
mayor wears the chain to symbolize his office on formal occasions.

The elected mayor who holds the office passes it on to the next
lord mayor at the end of each term. The chain is a token of elected
power, an immutable identity of typed authority. As an extension of
society, digital tokens confirming authority is critical. Capability-
based addressing implements immutable tokens to define specific
functionality. These golden tokens implement the digital gold of
Capability-based addressing, binding logic to physics., by naming
power and authority incrementally using immutable tokens to access
digital functions in the digital computer.

Using Capability-based addressing is the practical extension
of society into cyberspace. It avoids dictatorship and corruption,
distributing power democratically through function abstractions
and solving real-world problems using clockwork computer science.
Digital tokens are the gold of cyberspace, the digital treasure
that converts binary kingdoms into democratic institutions. The
comprehensible result is understandable by all, not just by experts.

The metallic strength of tokens of authority and power must
extend from the natural world into cyberspace. It makes everything
understandable. One quickly grasps how an abstraction of the City of
London as a namespace works. Membership is limited to legitimate

residents, each given a revokable token to access the London namespace as long as they remain London residents. The namespace holds annual elections among verified residents using agreed-on rules that run as smart contracts using blockchain technology when needed. Like beehives, the individuals share the same DNA as a colony that defines the roles and functions of the namespace. Notably, each settlement is self-standing, an independent, democratic unit without any branded central operating system.

Power is well and truly in the hands of the people, transparently freed from all opaque, dictatorial baggage. Individuals can ask to join any colony they wish. At the same time, the internally agreed institutions can decide whether or not to return a membership token, a digital capability token, to access the membership functions. Digital life is securely in the hands of individuals instead of dictators. Each colony is in total control of private processes programmed by the scientific laws held by users. The citizens are free and independent in democratic cyberspace, enforced equally and uniformly on the level computational surface of computer science, where it matters most.

Cyberspace is dependable, crime-free, and future-safe because the bottom-up binary details are bound to the top-down laws of the λ-calculus. These restrictions apply atomically and cellularly but certainly not monolithically. Necessarily, all share universal cyberspace, but the active namespace and a given thread define how far sharing goes and how resources compete. Nations can coexist independently as cyber colonies and define operating rules and government policies that evolve.

The two halves of the *Dream Machine* work in harness together. One side is a simple, single-algorithm Turing machine, reimplemented in the latest technology, leaving out all the baggage of centralized time-sharing. The other side is a Church machine with six λ-calculus instructions to control the namespace, the thread, the abstraction, the functions, the tokens, and binary data objects. These two machines respond to every machine instruction by cross-checking every action.

The combination pulls the individual threads as private computations within a namespace on behalf of individuals, working within the rules of an application namespace. These namespaces function like the DNA of a digital colony of individuals. Not all know and use every namespace. As in the natural world, an introduction links a contact that would otherwise remain an inaccessible secret. Following the introduction, an exchange of tokens links the two items together.

This simplified arrangement delivers a tremendous increase in performance in conjunction with digital privacy and data security. The distribution of parallel threads of execution and the delegation of golden tokens as secure function abstraction to local or remote parties result in less code and orders of magnitude improvement on every topic concerning computer science.

At the lowest level, the digital clockworks of *Dream Machines* understand the two internal machine types for binary data and Capability tokens. Tokens also customize access rights, making instruction type-safe, fail-safe, function-tight, and data-tight. Malware and accidental corruption are locked out. Blocking access attempts in advance allows for immediate recovery, creating industrial-strength software, and increasing the transparent powers of resource management that cover garbage collection, change control, fraud, and forgery.

The clockworks keep the software on the rails at high speed, including a measured mean time between failure (MTBF), just like hardware components. By knowing the MTBF, engineering modular software improvement is transparent, and progress is simplified because it evolves incrementally. The only significant alternative to exposed binary computers is Morello from ARM.[95] The architecture follows CHERI[96] and is a hybrid design that requires a centralized operating system. The improvement is, at best, a transitional step to the ultimate goal—removing centralization, binary compilations, backward compatibility, and the dictatorial operating system. Cyber society demands engineered software that these hybrid binary computers cannot achieve.

---

[95] Morello is a research program to radically change the way we design and program future processors and improve built-in security. Morello has a transformative goal to radically update the security foundations of the digital computing infrastructure of the entire global economy.

[96] CHERI (Capability Hardware Enhanced RISC Instructions) is a joint research project of SRI International and the University of Cambridge to revisit fundamental design choices in hardware and software to dramatically improve system security.

Capability-based addressing is now accepted as software technology, using tokens to abstract program links for software. The URL mechanism of the internet is an informal variation using mutable character strings that lack serious protection. Capability keys are immutable digital tokens, the gold standard of need-to-know security far superior and more potent in hardware than software technology. In hardware, tokens act as an exchangeable digital currency across cyberspace. They enable secure, functional programming to the hard edge of cyberspace, safeguarding access to everything in computer science, as intended by Alonzo Church. Capability-based addressing is the glue that holds the two halves of the *Dream Machine* together. It protects access to local memory and concurrent and remote function abstractions, including dynamically casting the immutable tokens for new namespace objects.

AI SOCIETY

The Information Explosion

Undetected software errors are bad enough, but industrial dictators undermine freedom and democracy. A democratic AI society must not, indeed, can not grow from this dictatorial binary foundation. The endless growth of cybercrimes means spies, criminals, and enemies forever dominate life in binary cyberspace. Preventing this gets harder and more expensive. Better skills to cope with solutions to binary attacks from different compilers running on multiple operating systems and using various binary formats only increase. Other dictatorial operating systems, each using an ever-expanding range of unfair privileges, options, and patches, attempt and fail to stop digital crimes.

The rules conflict, the picture is confused, the software is complicated, the binary code is opaque, and the breakdowns slow progress. At the same time, AI improvements push attacks to the limit, making patched upgrades more challenging and increasingly costly. But this dangerous binary concoction is an overstressed foundation still mindlessly trusted. Binary computations disconnect from mathematical science where and when it matters. Mindlessly trusting industrial dictators is a fatal error for AI society. They fail the tests of time and guarantee an AI breakout. Professor Max Tegmark at MIT[97] explains this catastrophe in his recent book that describes and illuminates the path-breaking advances in artificial intelligence that already approach human intelligence in language

---

[97] *Life 3.0: Being Human in the Age of Artificial Intelligence* by Max Tegmark.

and code production. Expect the crime rate to soar and threats to increase.

Only the flawless functionality of the Church-Turing thesis can resolve this dilemma; machines that apply mathematics rules flawlessly from bottom to top solve the difficulty. History proved that the clockworks of science powers progress, first for the abacus and later for the slide rule. Science is fair and democratically accessible to all. Science allows society to flourish and grow democratically. Pure science functions equally for all while detecting criminal attacks on the surface of the digital computer. Moreover, centralization evaporates when individuals use a *Dream Machine* to distribute privately threaded computations and replace von Neumann's centralized binary calculations.

The Church-Turing thesis defines the threaded rules of private scientific computations. Thus, *Dream Machines* replace monolithic centralization with functions encapsulated as digital abstractions under private ownership, engineering the flawless quality of computer science. Perfect, scientifically governed functional equality solves every cyber threat and empowers individuals and nations to reach advanced levels of democracy as progressive computer-aided civilizations. Indeed, culture only flourished when trusted arithmetic stretched along the Silk Road as the flawless mechanics of the abacus and industrial society only grew after the perfect slide rule. The ideal mechanics of a Church-Turing machine democratize the power of the λ-calculus. It protects AI society by replacing flawed binary computers with clockwork *Dream Machines*. AI society offers another level of civilized progress through scientific capabilities, transparently democratized for all to gain.

Sadly, today, the reverse is true. While good things get better, bad things only get worse. Undetected errors caused today by any source of corruption in binary computers undermine progress, increase costs, and create skill shortages. Digital crime and corruption inevitably harm industry, infrastructures, institutions, law and order, and democracy. Progress slows, and developments run late as highly skilled effort to maintain the software in the opaque, outdated, industry-frozen, binary computer only grows, detracting from new projects that improve society.

Backward compatibility deliberately stalled progress, and an exceptional effort is needed to support new and existing software on overstretched, opaque, and outdated computer designs. Compare this to Ada Lovelace's achievement. Once debugged, her program needs no maintenance. It is still a working expression of science implemented by mathematical machine code developed almost two hundred years ago. No other program of today can last that long. Monolithic compilation and binary machine code guarantee this. The cost of software development is too high. The result would be sensational if a *Dream Machine* doubled the current software's life by avoiding malware. Furthermore, development costs would fall dramatically, skill shortages would disappear, development costs would collapse, and Civilizations would progress.

The alternative is an industrial dictatorship that halts civilized democratic progress. The abacus, as a functional computer, won unanimous public support. The slide rule, likewise, took us to the moon. Why? Because they are trusted, transparent, and easily used, even unskilled citizens gain an individual advantage. No one needs a degree in obscure technology, branded operating systems, or multiple programming languages when the machine code works symbolically. It is self-explanatory. Anyone interested understands the declarative statements of λ-calculus populated by application-oriented names to define the specific functions of a problem-oriented namespace. To interested individuals, the terms are self-explanatory, and the machine code is readable.

Indeed, the key to the success of computer science is simplification and transparency for all to gain an advantage. It is the essential democratic characteristic of mathematical science; it exists uniformly for all to prosper and progress safely. Mathematical science is indispensable for cyberspace and democracy to thrive together and last forever, copying the staying power of Ada's Bernoulli abstraction.

To restate the most critical lesson, computers, software, and cyber society will only survive if innocent civilians can add to the power of globally networked computers as students. Smartphones, laptops, supercomputers, and desktops must defend, protect, and empower the individual against digital corruption. It is the only way AI society can avoid exploitation by digital criminals and dictators.

Trusted computer science is the engine of democratic civilization. Science won over the piles of stones that came first because engineered mechanics are trusted and accepted worldwide. The reliable power of the slide rule took humanity to the moon and beyond, once again, because results are dependable and trusted. In every case, the simplicity of the atomic nature makes machines worthwhile and transparent for all. But the opaque, monolithic, centralized von Neumann computer, running a Lampson authoritarian operating system, is outdated and dangerously different. These flawed, opaque, antique, and complex concoctions are not trusted machines. They even frustrate and confuse the experts.

Once again, civilization must solve the Babbage Conundrum, and the solution is certainly not a binary computer. Statically bound binary computers are piles of binary stones, another outdated approach abandoned in Babylon thousands of years ago.

Today, enormous human efforts limit corruption but fail to prevent cybercrime. Cybercrime becomes a Weapon of Mass Destruction. It is unacceptable for an AI society. Skilled staffing demands already exceed availability as hidden crimes grow and dictators tighten their grip and slow progress. Backward compatibility continues to stall progress, keeping human eyes closed to the Church-Turing thesis and the value of the λ-calculus. Dictators controlling cyberspace return to the past by enforcing outdated architecture and an authoritarian, centralized operating system.

They fear the disruptive clean break delivering Church-Turing machines and resist hardware improvements. Only ARM[98] has revisited a partial use of Capability-based addressing as a hybrid attempt following the work of the CHERI research. However, a total international commitment is needed to accelerate progress and simplify computer science, or the opaque, binary dictators will rule the world following the model of *1984*.

At the worst possible time, the progress of computers halted because operating system designers thought they could do it all, but clunky monolithic software development became so hard. The opaque nature of branded computer science drives corruption and an ever-increasing demand for highly skilled staff. But after decades

---

[98] WRM & Mongoos.

of arduous work, crimes still grow, forcing qualified staff shortages that slow progress. It is a downward spiral to a national disaster. At the same time, the Church-Turing thesis confirms the advantages of science, as already proven by PP250.

Still, industry dictators who run computer science question if a change is worthwhile. They have built their industrial dictatorships on outdated branded computers for over half a century. Disruptive progress is the last thing they want. Still, for civilization, progress is vital, and crime and dictatorship are unacceptable, so the answer will always remain a resounding yes to progress and no to backward compatibility. Adding the λ-calculus to the cockpit stretches science and security to the uniform atomic edge of cyberspace. It decomposes the centralized, monolithic virtual machines and the dictatorial operating systems into type-safe, fail-safe scientific abstractions.

In the same way the slide rule powered industrial expansion, the *Dream Machine* will power the burst of AI progress needed for cyber society to prosper without the risk of AI breakout. The wheel simplified the transportation of heavy goods, but the abacus democratized life globally by opening trade between strangers. Some used camel trains instead of carts and wheels. When the power of logarithms created the slide rule, the Industrial Revolution resulted in the atomic bomb, the jet age, the space age, and the advanced understanding of nature and biochemistry. Indeed, other things contributed, but the democratization of science through technology came first.

Lessons learned include the pioneers who proved the critical differences between pure mechanical computers that came first and the programmable computers that started during the Industrial Revolution. Charles Babbage's unfinished but flawless *thinking machine* could correct every error in the mindlessly trusted tabular data of the day.[99] But Ada Lovelace wrote and thoroughly documented the first recognized functional program. Programmable computer science using object-oriented machine code defines a new age of Civilized progress.

---

[99] Charles Babbage (1791–1871), mathematician, pored over astronomical tables calculated by hand. Finding error after error, he exclaimed, "I wish to God these calculations had been executed by steam." His appeal to machinery started a new era of scientific automation.

To prove and explain the machine, Ada programmed the Bernoulli function without a language compiler in just twenty-five mathematical machine statements, perhaps starting on a chalkboard as at school. Her programming language was no different than a teacher's symbolic statements of mathematics. For example, she explains in detail, in Note G, the interesting complex function that follows:

$$\int \frac{x^n \partial x}{\sqrt{a^2 - x^2}}$$

She went further with her Bernoulli numbers, easily readable by anyone interested, and like her example above, these programs stay as flawlessly true today as then, forever, and for everyone. These flawless mathematical statements of truth, computed by the powers of the λ-calculus, remain trusted and faithful. Mathematics is a science, unwaveringly true and perpetually correct. Software written this way is everlasting. The cost of code development and maintenance falls effectively to zero over a human lifetime.

The eternal nature of computer science is mathematics. Ada wrote her code a century before operating systems and virtual memory tarnished computer science. Programs written using the international language of mathematical symbols taught in schools worldwide remove all binary computers' opaque baggage, transparently democratize, and simplify computer science for every interested citizen. At the same time, it makes software safe from malware, dictators, and crime. When written mathematically, Ada's code and the code for a *Dream Machine* will last for all time, to the endless benefit of human society.

The positive results of a trusted *Dream Machine* on the future of civilization are beyond anyone's comprehension. Unleashing the full power of computer science using the Church-Turing thesis will enrich society in ways that are presently unknown. The flawless automation of functional fidelity will be infinitely productive, everlastingly future-safe, and universally easy to program. Scientifically safe, secure, transparent, and easy-to-use function abstractions will propel progress and, like Ada's examples, when done, will last forever without patches. Church-Turing machines guarantee the ultimate form of computer science, and the evolution of a stable, democratic AI society is quickly reached with less effort and cost because skilled staff shortages evaporate as progress accelerates.

Binary computers frozen decades ago truncated civilized progress because branded computers fragment cyberspace and enable cybercrime. Unavoidable, inordinate human effort is needed to keep programs on track. But science does a better job automatically and for free. All the lost effort is counterproductive. In the age of the mainframe, when industrial dictators froze the progress of computer science, the software was primitive and procedural, built for batch processing as corporate mainframes. Object-oriented programming and functional programs have changed the costs and time for software development, freeing the way to remove centralized operating systems. But while they remain, the essentials of the λ-calculus are blocked from crossing cyberspace. Backward compatibility must end for science to progress and find a civilized way forward.

The impenetrable, un-science of branded computers turns every mistake into a Mad Hatter's tea party. Unskilled civilians face increasingly opaque complexity. Too many unscientific mixed technology challenges are impossible to fix. Transparent science must replace the un-science of binary computers, and the opaque industrial dictators must fully adopt the Church-Turing thesis. Progress is moving in the wrong direction, undemocratic and silently corrupt. Powerful dictators make all the laws and place all investments in their interests, dictating the unpleasant future of nations.

Encouraging progress as software enhancements, languages, virtual machines, operating systems, and now AI while freezing hardware progress complicates, threatens, and fragments society, constraining constructive progress. Fragmenting cyberspace is unacceptable for a democratic society, and global progress is impossible because privileged operating systems digitally stand in the way. The effort to debug one compiled virtual machine is already too high, the time taken too long, and the result too problematic. Coordinating virtual machines across integrated cyberspace is impossible to consider. Rapid progress depends upon increased simplicity, not heightened complexity.

The Church-Turing thesis and corresponding Capability-based computers like PP250 transparently simplify and democratize computer science to pass every test of time for all. Free from crime and safe from dictators, transparent mathematics shines through on a level playing field for all to enjoy. After removing the dictator's finger from the scales, no unfair privileges or undetected errors exist. Transparent science is equal and fair, bereft of crime and unjust authority. Power exists equally for everyone, including amateurs and school children, to use to the maximum advantage.

Simplicity is the secret to the abacus, to the slide rule, in Ada's software for Babbage's *thinking machine,* and in the flawless computational, functional science of a digital *Dream Machine.* Functional modularity is the secret to trusted computers and the operational safety of AI society as an AI-enabled digital democracy. Banning backward compatibility will bring an end to cybercrime and cyberwars. Even more critical for the future, the superhuman powers of AI software will be enhanced by the constraints of digital modularity using the λ-calculus to prevent AI breakout. The AI-controlled deep fake[100] of George Orwell's Big Brother will be tamed as cybercrime is banished and AI is targeted correctly under human control.

As modular digital abstractions, functional software in *Dream Machines* can never exceed the purpose or intention. A λ-calculus[101] namespace has application-oriented guard rails to keep software, including AI software, on track at high speed. Guard rails prevent interfering hacks and block malware from entering the cockpit, simultaneously ending cybercrime and speeding results. The λ-calculus harnessed to Capability-based addressing to the very edge of cyberspace detects every digital error. Any programmed bugs left to discover cannot cause harm like a zero-day bug. The fail-safe

---

[100] Deepfake, Wikipedia, are synthetic media in which a person in an existing image or video is replaced with someone else. Deepfakes leverage powerful techniques from machine learning and artificial intelligence to manipulate or generate visual and audio content that can more easily deceive.

[101] The λ-calculus is a simple notation for functions and application. The key ideas are applying a function to an argument and forming functions by abstraction. The syntax is sparse, elegant, and focused on representing functions as rules of computational mathematics and logic. See the lambda calculus (Stanford Encyclopedia of Philosophy) for more.

Capability checks prevent this. They cannot impact more than one isolated thread of computation. Digital memory is always unharmed when an instruction trips over a boundary check, and corrections are easy and incremental, maintaining the total software investments.

Unlike the centralized dictatorship of a branded binary computer, the encapsulated hierarchies of each λ-calculus namespace blend seamlessly with the needs of a progressive democratic society. Consider a fail-safe namespace dedicated to the functionality of democracy, as discussed earlier. These tokens of democratic power function independently in cyberspace, like tokens in the natural world, shared by individuals for a limited time between elections.

The inauguration of a newly elected individual transfers power as tokens representing the robes of office from the old to a newly elected individual. In a Church-Turing machine, the robes of office are secret and secure access rights to particular function abstraction that formalize democracy in digital cyberspace. Given the flawless nature of fail-safe, tokenized software and the removal of unstable operating systems, no obstacles prevent national customs from framing independent forms of democracy. Big Brother's threats vanish when authority is organized and sustained democratically by and for all the people. AI society will progress safely for everyone living with endlessly improving democratic cyberspace.

THE FULCRUM OF CYBERSPACE

Fail-Safe AI Society

The seminal book *Understanding Media: The Extensions of Man*, written in 1964 by Marshall McLuhan (1911–1980), argues that the form of media changes civilization's development more than the content. Hence, the medium is the message. Today, the medium is computer science, interconnected as cyberspace. Consider, for example, how social media warps a child's development and how trust and lies change how adults perceive and understand politics, society, and the world. It is unclear if the information content polarizes humanity into today's warring clans or if individuals seek the support of extreme alternatives when hounded by cybercrime and disinformation. McLuhan argues the latter is most significant.

Further, McLuhan explained that all human inventions extend the individual. The human body, for mobility, uses the wheel to speed distances traveled, and the mind, through media, beginning with speech, then books written, printed, and transmitted, extend individual wisdom. McLuhan died before cyberspace formed and long before AI emerged. To explain the critical issues, consider that binary cyberspace, as used today, breaks McLuhan's rule. Individuals are suppressed instead of extended by dictatorial cyberspace. Deprived of control, individuals have no privacy. When aided by AI, this negative step brings about Orwellian society faster than appreciated. AI cyberspace is a weapon of mass destruction, with global obliteration powers exceeding the atomic bomb. Unchecked, the combination brings nations to unacceptable ends.

Freedom, equality, and justice are the cornerstones of a stable society operating as a democracy through individuals' willing consent and support. It is a virtuous cycle depending on the individuals ready to sacrifice when needed. However, individuals overwhelmed by vicious cyberspace cannot die for the cause because there is no way to fight when justice is already lost. Digital dictators and criminals rule cyberspace; individuals have no rights. So, danger looms, crime wins, governments spy, and fear replaces optimism. Individuals succumb to extremists because individuality evaporates. Vital support for democracy evaporates along with freedom, equality, and justice. Binary computers are killing society.

For AI society to prosper, democracy must exist in cyberspace and foster individuality over dictatorship. Computers must support individuals and reject dictators. Only actual computer science provides the answers. The powers of individuals must be systemic for liberty, equality, freedom, and justice to prosper in cyberspace and allow AI society to evolve democratically and accelerate human progress instead of war. As clarified by McLuhan, democracy must emanate as the message of cyberspace by extending the power of individuals. Then, and only then, can AI society prosper and grow as an evolving democratic cyber society. Cyberspace is an existential threat when control is centralized. Binary cyberspace crushes helpless individuals with opaque complexity, fear of undetected crimes, and dread of digital spies and government henchmen, the industry dictators of the binary computer.

Inevitably, individuals turn to tribes and cults to secure their future. Instability results in the West, while Orwellian dictatorship grows in the East. Unless binary computers are banned, dictatorship will take over the world. Prohibiting the sale of new gas-powered vehicles addresses Global Warming, so prohibiting the sale of outdated, dangerous binary computers would deal with another equally serious WMD. A weapon of suppression by centralized power systems, aided by global digital technology, deep fakes, pervasive cameras, and easy cybercrime, guarantees an Orwellian society. It is approaching fast. Polarization and tribalism are the early warning signs of an impending and unavoidable national catastrophe.

The experts run cyberspace for themselves. They toy with superuser powers to increase control, crush individual freedom, equality, and justice, steal their privacy, publicly share secrets, eavesdrop, spy, rewrite history, and enslave users. Colluding industrialists, criminals, and government tyrants are the warlords of uncivilized cyberspace. McLuhan was right; the present message from cyberspace is loud and clear: crime wins, and individuals mean nothing. Results will worsen, and as innocent citizens surrender, ending democracy as polarization makes everyday life intolerable and governments turn to dictatorship. Unstoppable international wars will then impoverish AI society.

Industrial dictators define the powers of the binary computer to stay in the lead. Backward compatibility undermines progress and polarizes society, but it makes dictators rich. It is the primitive procedural code that brings them rewards at the expense of citizens. Centralized binary computers offer no rights of individual privacy or security to defend their independence in cyberspace. Everything hinges on dangerous centralized sharing that began when hardware was expensive, but centralization is a technology for dictators and criminals, not for AI society. Ignoring the Church-Turing thesis and the λ-calculus is an unforgivable error of computer scientists now threatening the future of civilized progress.

Government action must replace the flawed binary computer—a combination of urgent regulation and well-considered investments—another Manhattan project to 'Decentralized Computer Science' by following the science of the Church Turing thesis. Symbolic machine code is the fulcrum around which everything turns. Machine code is the lowest but the most significant of all programming languages.

Binary machine code formats in every existing computer, using shared physical addressing instead of symbolic addressing, must be banned from sale within a decade. Only this will prove the industry's commitment to fixing malware for good. The power of functional programs must interwork through pure mathematics and logic that underpins computer science into the everlasting future. In absolute terms, machine code functionality defines digital computer science through functional abstractions calibrated to compete on quality, performance, reliability, and security. Powers inherited from the surface of computer hardware crosscheck life in two separate ways. The complementary registers of a *Dream Machine* bring fault-tolerant software to life as an individually controlled computational extension of democracy.

Step by step, the machine code of a future safe cyber utility expends electrical energy as programs run. This digital activity drives attached hardware devices to improve privacy and increase prosperity. The tokens represent incremental power distributed fairly to private threads. The threads remain confidential because they belong to individuals concerned about family life and future survival who understand personal secrets. The security of the computational thread is in the hands of each user as they journey through cyberspace. The fail-safe machine code must detect every error and keep malware out of the cockpit. The guard rails keep every digital action on the right track.

Undetected errors cannot exist, blind trust must disappear, and centralization must vanish from the architecture. The opportunity to find and fix the mistakes is performed mechanically through the clockwork machine code. This computer cannot take dangerous shortcuts. It must allow students and young children to write mathematical machine code safely for themselves and all others. The talisman of success uses the science of the λ-calculus to recreate an Ada Lovelace for the future. Competition using the Bernoulli function to solve the Babbage Conundrum will distinguish the best machines for various tasks.

Alonzo Church's computational model transparently solves every issue from scientific first principles. The dynamic scientific result of functional, abstract computational modularity, defined by symbols of the λ-calculus, establishes the opportunity for Flawless Distributed Computer Science. Fail-safe, future-safe technology protects individuals and targets society to democracy instead of dictatorship across cyberspace. The functionality of the computer's machine code drives all this.

Consider the implications for the future. Instead of crime and dictators, science and individuals will define the future. The machine code directly executed by the hardware on the surface of computer science is the lowest level of cyberspace where science and security coexist. It is not and should not define a monolithic virtual machine. Doing so requires a centralized operating system when, once again, we all fall down von Neuman's rabbit hole. Instead, the machine code defines a scientific machine with physical limitations—functional machine code for the real world. The imagined world is the language of function abstractions. The scientific machine code as function abstraction is the language that materializes cyberspace by preventing crime and centralization. No special instructions can exist that require a superuser. Instead, equality and individuality are the touchstones of success. Removing the superuser prevents the threats of ransomware. Mathematics and logic resolve the dangerous inequality that leads to dictatorship.

The task of programming machine code is called assembly. It involves generating individual binary statements called machine code. As a Public Utility following the PP250 example, no longer done by hand, an application-oriented command interpreter is built-in. As a future-safe, fail-safe computer, the *Dream Machine* includes an IDE (Integrated Development Environment) to help write declarative machine code statements and keep them on track for eternity. The namespace defines secure, readable functionality without a compiler or an operating system. Application-oriented languages to improve productivity derive from the abstractions implemented by the namespace.

Looking back in time, almost two hundred years, to when Ada Lovelace wrote the first program, she addressed the question symbolically. Until then, computers were purely mechanical, but Charles Babbage proposed the programmable alternative called the Analytical Engine, directly using mathematics and logic to define the machine instructions. Ada's program is easy to read and understand because it begins with symbolic mathematical expressions. This form of machine code is the IDE taught to kids. It is universal. It never ages and constantly improves. Ada's code is eternal and should be so for schoolchildren and cyberspace. Using λ-calculus to guard and guide the object-oriented machine code stretches science uniformly and globally across cyberspace without any cracks or gaps needed by malware and without any centralized dictators. This scientific implementation of cyberspace is flawless. Individuals define the computational threads computers use as private actions, the same way privacy and security exist with the abacus and the slide rule.

It is the exact mechanism enforced by the λ-calculus. It is the perfection of computer science as defined by the Church-Turing thesis.

By encapsulating and thus hiding the exposed and volatile binary engine within a λ-calculus expression, the function abstractions of each natural namespace become real. The Church-Turing thesis simplifies, secures, and democratizes computers and programs as transparent and reusable functions that skilled and unskilled citizens can treat as a black box. Furthermore, the vital benefits of readability, privacy, security, functionality, productivity, and reliability improve functional performance, automatically built into the machine code with tokens that can reach throughout the known digital universe. Its simplified distributed model removes the monolithic complexity of opaque, incompatible binary computers, dissenting compilers, and an unfair superuser operating system.

The inherent simplicity of Turing's alpha machine is preserved and protected by removing all the opaque baggage from decades of attempts to salvage von Neumann's shared binary architecture. One algorithm runs in one abstraction at a time using a limited list of named relationships. The list associates the related digital objects and contacts foreign objects accessed by the need-to-know secrecy of top security systems.

The PP250 successfully demonstrated this computational model decades ago; word lengths have changed, and improvements are needed. Still, as the first networked computer architecture, it established the demanding requirements for fault-tolerant hardware and fail-safe software. PP250 enforced the laws of the Church-Turing thesis using a machine with only twenty-four bits and achieved decades of fail-safe software reliability. The future of global cyberspace will dramatically improve if regulated as a public utility. This single step brings sanity to cyberspace.

ARM has kindled progress with its hybrid solution, but a more profound step cuts the cord and walks away from von Neumann's dangerous mistake. Only a Church-Turing machine empowers the democratic survival of AI-enabled cyber society. The dream depends on mathematical perfection and privacy in cyberspace, a computer system that is fail-safe and proven so. The *Dream Machine* is flawless, fail-safe, and easily understood by anyone interested and familiar with basic mathematics. Once written and tested, these programs last. As proven by Ada Lovelace, mathematical programming lasts unpatched forever. Each of these goals for the ideal computer is verified. Mathematics and logic are, by definition, flawless. Therefore, computer science correctly engineered follows suit. Transparent computers were born at the birth of civilization, while the redoubtable slide rule enabled the space age. Even more critically, Ada Lovelace and Charles Babbage proved a flawless functional computer programmed mathematically using the language of science that lasts forever.

Invariably, digital corruption emerges as binary computers run. Harmful programs are inadequately tested or deliberately crafted as malware. Accidentally and deliberately, AI malware finds gaps in the digital void of monolithic centralization where virtual machines run using default privileges exposed by hardware and software. In every case, the result is machine commands with unfair default powers beyond the scope of the required context. This condition is unnoticed but corrupts the delicately balanced configuration of a binary computer. Undetected corruption lies dormant from when it occurs until discovered later by accident or as a digital crash that can cause a catastrophic humanitarian disaster.

When binary computers began after World War II, the experts thought a handful would meet worldwide needs. They were wrong and failed to understand the Church-Turing thesis or the technology of Capability-based addressing. Then the ratio between skilled staff and binary computers was high. Large information technology departments kept the software on the rails by limiting all changes and isolating critical services in locked rooms. Networking became a back door used by remote hackers and remains dangerously under guarded.

Hacking soon became a profession founded on corruption and the root cause of cybercrime. In the days of stand-alone batch processing, large information technology departments tolerated the binary flaws by preventing untrusted connections, slowing the change rate, and only working in eight-hour shifts. It took months of arduous work, and when batches did not run correctly, they ran again on another day. Slowing progress is the same strategy used today that froze computer development, preventing the adoption of Capability-based addressing. Suppliers use their branded designs to hang onto markets despite the damage done by criminals and enemies to society and nations.

The mainframes ran programs batched into eight-hour shifts, with printed results spewed out by massive line printers. Finding errors was up to users checking printouts. Shifts always restarted from the factory default condition that purged the memory of any digital garbage and programmed malware.

Since then, driven by microprocessors and personal computing, the software has shifted from isolated programs into a global network of endlessly working, nonstop, parallel computations and distributed calculations. Multiprogramming began with the SABRE[102] airline reservation system in the early 1960s and soon after that with fault-tolerant, real-time telecommunications[103] (AT&T 1ESS). The regulated telecommunication industry demanded standards for public service the mainframe never faced. Networked telecommunication software must work, day after day, for decades. As a multivendor public utility like the telecommunication industry, only the highest standards serve the public interest—standards the computer industry ignores.

Shifts and breakdowns familiar to mainframes and minicomputers are unacceptable for public service. Indeed, for anything other than mainframes. However, despite research started at MIT by Jack Dennis on changes needed to support multiprogramming, nothing changed. Even Multics at MIT ignored Capability-based addressing. The meta-instruction for parallel, distributed, thread-based computations[104] and the search for flawless software fell to the telecommunication industry. The highest standards of public service forced this level of achievement.

---

[102] In 1964, Sabre's nationwide network was completed and became the largest commercial, real-time data-processing system in the world. Sabre Corporation handled 7,500 passenger reservations per hour in 1965. The Sabre system upgraded to IBM System/360 and moved to a new center in Tulsa, Oklahoma, in 1972.

[103] The Number One Electronic Switching System (1ESS) was the first large-scale stored program control (SPC) telephone exchange or electronic switching system in the Bell System. It was manufactured by Western Electric and first placed into service in Succasunna, New Jersey, in May 1965. The switching fabric was composed of a reed relay matrix controlled by wire spring relays, which in turn were controlled by a central processing unit (CPU).

[104] Review: Jack B. Dennis and Earl C. Van Horn, Programming Semantics for Multiprogrammed Computations. See https://courses.cs.washington.edu/courses/csep551/04wi/Messages/paper3.

The λ-calculus avoids centralization, allowing individual threads to use commands like *thread.fork, thread.quit,* or *thread. join* programmed directly without an operating system. These scheduling functions support parallel processing through abstraction, specifically for this example, by abstracting a computational thread of execution that exposes the functions *fork, join,* and *quit.* The centralized operating system vanishes. For computer suppliers, there are downsides in disrupting their markets, so encouraged by flawed claims about centralized operating systems, computer science stayed stubbornly frozen in the 1960s.

Only UK telecommunications and the British Armed Forces used Capability-based addressing for a decade after 1975. But the dangers of cybercrime still grow, international cyberwars expand, and as a result, revisiting Capability-based addressing has taken place. For example, the paper Capabilities Revisited: A Holistic Approach to Bottom-to-Top Assurance of Trustworthy Systems by Peter G. Neumann[105] at SRI and Robert N. M. Watson[106] at Cambridge Computer Laboratory published in 2010 began the CHERI research program and the resurgence of interest in Capability-based addressing. It led to ARM's Morello Computer. This transitional computer is a hybrid design that still depends on a privileged central operating system. However, it marks a significant acceptance by the computer industry that might overcome disinterest in Capability-based addressing. Capability addressing uses immutable names as tokens that define access rights in cyberspace. Tokens are the foundation technology for flawless computer science built into the Church-Turing thesis, λ-calculus, capability addressing, and object-

---

[105] Peter Gabriel Neumann (born 1932) is a computer science researcher who worked on the Multics operating system in the 1960s. He edits the RISKS Digest columns for ACM Software Engineering Notes and Communications of the ACM. He founded ACM SIGSOFT and is a fellow of the ACM, IEEE, and AAAS and advocates CHERI research as a transformational technology for secure, trusted computers.

[106] Robert N. M. Watson leads research projects at Cambridge University on automated software analysis and decomposition for security, revisions to the hardware-software interface for security, and a project in novel hardware and software designs for secure cloud computing. Watson proved that real attacks can be developed on the current multicore central processing units (CPUs) and showed that developers must fix their code.

oriented programming. When cyberspace becomes a government-recognized public utility, the enforced standards will drive progress forward. To achieve decades of software reliability with object-oriented machine code and mathematics that lasts forever. Computer science must adopt the science of Church and Turing to secure digital modularity in AI-enabled cyberspace.

Flawless computers are not new. They began at the birth of civilization and culminated with Charles Babbage. He perfected his idea of the *thinking machine* almost two hundred years ago. His clockworks keep functional computations on track as pure mathematical calculations, proving beyond doubt the impact of perfected computer science on civilized progress, society, and democracy.

Programs must stay on track by enforcing the logical context of each computational step. Capability-based addressing is the hardware technology used to execute the computational laws of the λ-calculus. Capability registers define the unlocked local context for machine instruction to detect errors, including AI malware breakout attacks and every undiscovered program bug. The immediate discovery of any error in a single computational thread prevents digital corruption and makes quick recovery easy. Rather than worry most about system-wide threats, service failures, file corruption, and deliberate interference, everything necessary simplifies to fail-safe, data-tight, and function-tight machine code. It is another case of caring for the pennies while the pounds care for themselves.[107]

Name-based, dynamic binding detects every error as a neutral, independent judge, regardless of the intent. Both accidental and deliberate attacks are detected and prevented by dynamic binding the atomic actions of every machine instruction. It is how PP250 achieved long-term software reliability and why the abacus has survived for thousands of years. It is also why Ada Lovelace's program will be an eternal landmark for programmers to replicate and, like her example, last forever without needing a patch or a rewrite. Capability-based addressing is the technology to perform the digital tasks that bind the Turing machine as the λ in the λ-calculus.

---

[107] William Lowndes, the British secretary of the treasury, 1696–1724.

Consider a decimal number as a λ-calculus function abstraction programmed for a *Dream Machine*. For this example, a decimal abstraction performs the following expressions. *SET*(x) and *GET*() are used to initialize and retrieve the stored number intelligently. The function abstraction needs just two other modes as functional methods *ADD*(x) and *SUBTRACT*(x), where x is the second number. At the atomic level, one rail of an abacus is a number x between zero and nine, while the result could reach eighteen. The exact implementation details are of minor concern to the user if reliability is qualified with an MTBF. The abacus scales and works reliably for all conditions because the atomic implementation is secure. The rails and frameworks achieve this goal. In computer science, the λ-calculus does the same when implemented by Capability-based addressing for any scientific function abstraction, regardless of the power of the expression. It works for AI in all forms.

The decimal number is a type. Types exist to support other functions, as with a slide rule. As a digital black box, an instance of function abstraction is always a number. A formal classification of digital implementation represents each one. This individuality is the nub of the difference between a binary computer that is digitally unstructured and a Church-Turing machine. The λ-calculus structures information functionally and atomically, and Capability-based addressing guarantees the details through lambda machine code.

Finally, and perhaps most importantly, if McLuhan and history are correct, computer science as a power source should extend the individual, not result in authoritarian governments. More definitely, an AI-enabled society must enhance children as well as adults. It is a subject of individual rights that must be equally shared by all, not hoarded by industrialists, gangs, governments, and enemies. The Rights of Individuals in cyberspace is a national concern driven by the written Constitution of the United States. The freedom of society as individuals must be protected in the virtual world of cyberspace to remain in the real world. Flawless computer science is the only option for AI-enabled civilization. America promises every citizen freedom, equality, and the pursuit of happiness. America needs the digital *Dream Machine* to sustain individual liberty and independence. Safe, secure digital forms of life where everyone is free from undetected interference, accidental errors, malware, dictators, and deliberate corruption.

Industrial suppliers must change and support individual rights in democratic cyberspace and push forward safely and secure digital frameworks for democracies as AI-enabled societies. It is unavoidable, but if the governments fail to enforce these standards for the public good, human greed, short-term profit, and unpunished criminals will destroy AI society. Then, shattered civilizations will restart the painful search for a *Dream Machine*. It will rise like a phoenix after the unavoidable binary computer cyber catastrophe.

KEN HAMER-HODGES FIEE

Author

Ken Hamer-Hodges paid his way through college and joined Plessey Automation in 1966 as a junior engineer, designing memory units for IBM computers.

At that time, the world was still mechanically powered. Electronics, computers, and software were DIY projects. A memory system of ferrite cores as big as a large refrigerator is now on a semiconductor chip with the computer included. Telecommunication factories still made mechanical switching technology powered by rotary phones, a patent granted to Almon B. Strowger in 1892. I soon joined a research team investigating computerized telecommunications. We started from nothing with the one overriding requirement to meet the public service standard of five decades of reliable public service without a crash. Computers were not yet commercial products; software was homegrown, and the reliability goal was severe. Professor Sir Maurice Wilkes from Cambridge University came to the rescue, supplying papers, research work, and introduction to those studying Capability-based addressing. He changed my life and gave me a vision of reliable computers and software as engineered science.

I led the microprogramming team at Plessey for the first commercial Capability-based addressing computer, PP250. We developed and used object-oriented machine code that enabled functional programming in machine code. Although the full measure of the Church-Turing thesis and λ-calculus was unrecognized, we had already decided that Capability-based addressing and object-oriented programming were fundamental to engineering a reliable digital solution. When the computer industry did not accept this truth and microprocessors changed the world, Capability-based addressing went to sleep. But PP250's object-oriented implementations went ahead, thanks to my friends Brad Cox and Tom Love, who sold Objective-C to NeXT and Steve Jobs. Steve proved his vision by adopting Objective-C, which led to Apple's resurgence from near bankruptcy to the worldwide leader in friendly computers and smartphones following the iPhone.

Object-oriented software took time to catch on, but now, many such languages exist in all sorts of variations. However, the magic of PP250 came from significant productivity gains using functional programming. With the λ-calculus and Capability-based addressing, a Church-Turing machine starts a new age of computer science. An independent λ-calculus namespace with decades of software reliability is the future's software foundation for democratic civilization. Nothing else will fill the dream. I try to explain how a *Dream Machine* brings together everything I learned along the way, hoping the dream will come true one day. If not, I fear AI society will end catastrophically. Civilized society cannot endure von Neumann's binary computer running global cyberspace.

CIVILIZING CYBERSPACE

The Fight for Digital Democracy

The subject of cybersecurity is too dry, while a readable text is too short. Instead, I hope to provoke a loud debate on the future and the urgent need for industrial-strength computer science. Everyone is involved; damaging cyber society by accident in the pre-electronic age was never cataclysmic. However, the critical flaws in the vital technology of the electronic age will destroy twenty-first-century civilized society. The general-purpose computer is the enemy. It changes the grand experiment of a free society and American justice. Super-intelligent AI malware cannot run law and order in Life 3.0 if it roams wild. Cyber democracy only exists if citizen users hold the keys to privacy and security in cyberspace. Within a decade,

AI-enabled software automation in every conceivable form will dominate life in the globally interconnected electronic village of the twenty-first century. However, the global village uses the pre-electronic age general-purpose computers. But for cyber society to survive, trusted computer science must exist.

A Church-Turing machine responds to this challenge; it is the proven way to guarantee trusted software throughout cyberspace, using an electronic age architecture from the Church-Turing thesis. General-purpose computer science began in the pre-electronic age of WWII and the Cold War. It grew like Jekyll and Hyde into an international weapon of mass destruction disguised as a recreational stimulant. Through digital convergence, this WMD is already in the hands of envious criminals and hostile enemies. The added threat of a breakout by super-intelligent AI malware creates scenarios too horrific to contemplate. The point of singularity is near. Artificially intelligent malware is both unspeakable and unstoppable. It is already in use, and a solution is needed urgently. Industrial-strength computer science as Church-Turing machines has become a national priority.

Printed in the United States
by Baker & Taylor Publisher Services